Table of Contents

Introduction 7

Part I: The Story of a Girl Who Constructed Her World

FAITH

Distress
    The Trouble with Third Grade—Britt 10
    Third-Grade Mother—Julie 14
    Long Long Road—Carol 17
    The Beginning: Dr. Medlar's Office—Julie 20

Assessment
    Meeting Britt—Carol 21
    Meeting Carol—Julie 22
    Meeting Julie—Carol 24
    Struggling to Understand—Britt 27
    Why Are Easy Things So Hard?—Carol 31
    Fourth Grade—Julie 34
    Visiting Britt's School—Carol 37
    The Rest of the Picture—Britt 42
    Why Are Hard Things So Easy?—Carol 46
    Daily Life—Julie 48
    It's Not Sticking—Carol 49

Answers
    Finding Answers—Carol 52
    Having Answers—Julie 55

HOPE

Space and Spatial
    Inner Space Is Not Outer Space—Carol 58

Constructions Class
    A Window Is for Seeing—Carol..61
    The End of Our Rope—Julie...64
    "It's Good for You"—Britt 67
    Planning the Constructions Class—Carol 68
    Did I Waste My Money?—Julie...70
    Summer School—Britt 71
    Beads and Blocks—Carol 75
    A Window Joins More Beads and Blocks—Carol 78

Table of Contents, continued

    Discovering Shapes—Britt 80
    Learning About Learning—Julie 84

Going Forward
    Ms. Jenkins' Class—Britt 86
    Someone in the Kitchen with Mom—Julie 87
    Pancakes, Cookies, and Pies—Britt 89
    Things I Learned in Middle School—Britt 90
    Middle School Academics—Carol 93
    Middle School—Julie 94

ACTION

High School
    Transition to High School—Julie 98
    Getting Ready—Britt 100
    Annie Wright—Britt 102
    Cracking Algebra's Code—Carol 107
    Life Beyond the Classroom—Britt 111
    Teaching Britt to Drive—Julie 113
    Learning to Drive—Britt 115
    Driving *Is* Spatial—Carol 118
    Commuting with Britt—Julie 121
    On the Road—Britt 123
    Geometric Shapes—Britt 124
    Making Geometry Happen—Julie 126
    What's Your Angle?—Carol 129
    Moving Forward—Britt 133
    Working—Britt 137
    Preparing for College—Julie 139
    Finishing High School—Britt 142
    Why Did Hard Things Get Easy?—Carol 145

College
    Going East—Britt 148
    Parenting a College Student—Julie 152
    Finding My Place—Britt 155
    Parent Advocacy—Julie 158
    Graduation—Carol, Julie, and Britt 159

Part II: Helping Students with Visual-Spatial Problems

    Background 163
    Getting Started 164
    The Language of Space... 166
    Grammar and Language Mechanics 170
    Learning to Tell Time.. 172
    Mapping 174
    Handwriting and Drawing... 176
    Distance, Weight, and Size 178
    Mathematics
        Number and Place Value 179
        Operations 181
        Story Problems 183
    Writing.. 185

Selected Bibliography: Spatial Relationships 187

Introduction

The Source for Visual-Spatial Disorders tells the story of a girl who learned to deal with a visual-spatial disorder. It also provides practical advice about how to teach those students who are struggling with a visual-spatial disorder.

Part I

In *Part I: The Story of a Girl Who Constructed Her World*, three authors tell the story of how a child learned to overcome the problems connected with a spatial learning disorder. Britt, the girl with the spatial disorder, relates how it feels to be nine and able to enjoy *Little House on the Prairie*, a book written for much older readers, while at the same time unable to subtract 1 from 7. Now a graduate of Smith College, Britt describes the pain and the joy of her journey.

Julie, her mother, who teaches at the University of Puget Sound, tells how she enjoyed her daughter's precocious language as Britt chatted in sentences at a year old. Later, during Britt's early elementary school years, they pondered plot and character as they shared reading aloud. Julie was puzzled that the same child who memorized pages of poetry seemed unable to recall 4 + 5 = 9, or to recognize that the answer was the same as 5 + 4.

Carol, director of a learning clinic, discerned a pattern in Britt's problems. She became intrigued by spatial disorders, by how invisible the condition was to most people, and by how often Britt was blamed for her inconsistent performance. She sought to understand the condition by studying research in medicine and perception. Based on this study, Carol planned instruction designed to build alternative foundations for understanding space. The activities that Britt carried out were then transposed into language. In the process, spaces that had been invisible became visible. But first came faith.

Faith, the first of three sections within *Part I*, tells of discovering Britt's medical condition along with the realization that it could not be altered. The discovery began with a growing awareness of just what this disorder meant for the mother, the child, and the teacher. Together and separately, they confronted reality by defining the condition, discounting the negative and doubtful, and proceeding with the faith that Britt could succeed.

Hope, the second section in *Part I*, portrays Britt's tenacity, the mother's resolute courage, and the teacher's inquisitive search. They discovered the practical implications of "spatial disorder" and persisted even when persistence required that they use unconventional methods to go forward.

Action, the last section in *Part I*, describes Britt as a capable adolescent and young adult whose literary interests led her through a challenging high school and college curriculum with her mother's support and Carol's decreasing help. Britt gently but actively informed the unaware teacher or college president when she was put at a disadvantage. She did not tolerate the intentional discrimination although her smile and wry humor often made the offender wish to be fair. She learned to drive! Britt, who became lost one block from home, now navigates the freeways of Los Angeles.

Introduction, continued

Part II

Part II: Helping Students with Spatial Problems provides concrete suggestions for remediation of spatial problems in a variety of areas. Although teachers and parents will have to deal with problems specific to the individual, the authors hope that these techniques will help those dealing with a visual-spatial disorder develop effective strategies for living and learning.

- Space
- Grammar and Language Mechanics
- Telling Time
- Mapping
- Handwriting and Drawing
- Distance, Weight, and Size
- Mathematics
    - Number and Place Value
    - Operations
    - Story Problems
- Written Composition

Conclusion

Many people equate learning disabilities with dyslexia, a disorder centered in the language system. Britt does not have dyslexia. She has difficulty perceiving space and spatial relationships despite 20/20 eyesight. Also called a non-verbal learning disability, a spatial disorder may result from a variety of causes. Britt's problems were caused by a medical condition called Turner's Syndrome, but other causes include restricted oxygen to the fetus during the birth process and events after birth such as illness, head injury, stroke, or near drowning involving oxygen loss to the brain.

The Source for Visual-Spatial Disorders shows how much you rely upon perceiving space and what the result would be if you could not. Routinely, people make judgments that depend upon knowing how near, how far, how much, how long, and where. Britt, despite being highly intelligent, did not "know" these spatial details. You may know people who have learning disabilities similar to Britt's. This book will help you appreciate both the weight of the burden and the solutions that are possible.

The Source for Visual-Spatial Disorders portrays the route anyone with a spatial disorder must take to meet personal goals. All achievement comes at a price. Whatever the obstacles, the challenge is to find a way and to enjoy the journey. We invite you to join Britt, Julie, and Carol on their journey to understand how to cope with a spatial disorder. Our mission is to show how faith allowed hope sufficient room to grow. Hope, in turn, generated methods and attitudes that became Britt's foundation for an active and assertive life. In the process, she created the space to build her own success.

We hope that with the help of this book, you and your students will learn successful new ways to grow and construct space.

Carol, Julie, and Britt

Part I:
The Story of a Girl Who Constructed Her World

Faith

*"It is at night
that faith in light
is admirable."*

Edmond Rostand
1868-1918

Britt: The Trouble with Third Grade

Sitting here straining to remember the years I spent relearning almost everything spatially oriented and fighting against the expectation that I would never attend college, I am forced to view these recollections as an opportunity to share my experiences with others. Like so many people with learning disabilities, I found there are many moments when <u>fear, panic, and feelings of isolation</u> become all consuming and overwhelming.

Third grade was the first time anyone recognized my struggle with school. Even though I had high grades in reading, social studies, vocabulary, and music, I struggled with math and spelling. My teacher in third grade, Mr. Palmer, noticed a discrepancy between my abilities in reading and language and my difficulties with math and spelling.

As a ten-year-old, I tackled simple math problems, mapping assignments, spelling tests, and copying projects valiantly. Still, no matter how hard I tried to please my teachers, I found it difficult to end up with the right answers. Although unaware of the cause, I knew that school had become difficult. Teachers told me that with extra effort, my grades would improve and that I just needed to try harder. With this in mind, I studied simple addition and subtraction problems at the kitchen counter every afternoon until my eyes began to burn.

During the summer before third grade, I learned that Mr. Palmer, a family friend, was going to be my third-grade teacher, so I approached the school year with enthusiasm.

When I first walked into the classroom, I wasn't disappointed. A light and cheery place, the room was a rectangular shape with a wall full of windows looking on to the front lawn of the school. The alphabet, written in cursive, formed a border as it paraded around the room, and a large bulletin board in the back displayed an array of brightly colored autumn leaves cut from construction paper and neatly placed around a brown tree stem. Clean and neat, the aroma of chalk and paste mingled with the early fall breeze drifting from a half-open window. The students' desks paralleled the windowed wall facing the teacher's desk and chalkboard at the front of the room.

Just before the morning bell rang, I hopped into my chair that was, as usual, too big. Although the desk was built for third graders, my legs dangled from the chair.

Mr. Palmer, about thirty with a neatly trimmed mustache and keen smile, was a favorite among all the students at my elementary school. His innovative teaching methods created a dynamic, energy-filled classroom that still inspires happy memories for many of his former pupils.

Math assignments often turned into games and competitions. In the early spring, Mr. Palmer encouraged us to learn multiplication tables with a contest: a wonderful idea if math is your strong point.

Quieting the class, Mr. Palmer explained the next math assignment in detail.

"Now that most of us have done all of the addition and subtraction problems in the math workbook, it's time to move on to multiplication tables." A general murmur erupted from the students.

"Not all of you are ready for this challenge yet," Mr. Palmer said, looking straight at me. As I lowered my head to avert the glare, I felt the blood rush to my face. Halfway through the year, I had still not mastered borrowing or carrying in subtraction and addition. I was

The Source for Visual-Spatial Disorders

Britt: The Trouble with Third Grade, continued

lost without a piece of paper or number line to calculate *nine plus four* or *eight plus two*.

"But most of you need to move ahead. How many of you know how to multiply?" he asked. Several students raised their hands eagerly hoping to impress Mr. Palmer with their answer. I sat with my head still lowered.

"I've made a board with everyone's name listed down the left-hand side and multiplication tables listed at the top, starting from one and ending with ten. We're going to use this board to keep track of each student's progress throughout the next few weeks. Each time you successfully pass a multiplication table, you'll receive a star next to your name showing how many tests you've passed. At the end of the year, we'll have a picnic and campfire on the beach at my house for all of you who pass the multiplication tests. Does anyone have questions?"

A general buzz erupted from the classroom. Hands flew straight up in the air along with waving arms and loud, enthusiastic voices.

"Will we have hamburgers or hot dogs at the picnic?" I heard one of the kids shout over the general noise emanating from the sea of desks.

"I'm not sure yet. Let's take a vote closer to the picnic," Mr. Palmer replied. "How about s'mores?" another excited kid quizzed. "Maybe, we'll talk about the menu a little later. Right now I'm going to hand out the papers with the first set of multiplication problems."

Day after day, I struggled to finish the multiplication tables. I practiced at home every night, reciting the tables perfectly aloud, but when I sat down at

> **" Feeling paralyzed, I stared at the paper hoping that the answers would magically come to me. "**

my desk with the test sheet in front of me, I could not keep the numbers straight. Making up rhymes to remember the times tables helped in memorizing, but it did not help me to understand that 5 x 6 is exactly the same as 6 x 5.

I walked into class every morning ready to take on any math problems that Mr. Palmer threw at me. Each time I told myself that this time, this test would be different. I'd pass the test on the first try if I concentrated hard enough.

But, when I had the page of math problems in my hand, I felt uneasy with my pep talk. Feeling paralyzed, I stared at the paper hoping that the answers would magically come to me. Realizing that time was important, I attempted the first problem unsuccessfully. Frustrated but undaunted, I tried the second problem and again failed to answer the question correctly. With the third problem, I knew that I had the answer right, but I transposed the number when I wrote it down on the test sheet. Frustrated, I forced myself to work through the rest of the test. Ashamed, I handed the test to Mr. Palmer, who took the sheet and shook his head disapprovingly as he made red marks next to all but two of the ten problems.

"Britt, you'll have to do better than this to pass," Mr. Palmer sighed, as he handed me the test sheet doused with red marks.

"I know, I'll try again tomorrow. Maybe I can pass if I practice more." Disheartened, I slowly made my way back to my seat.

Realizing that I needed extra help, Mr. Palmer set up tutorial sessions. Every afternoon I was excused from art or music and sent down the hall to the library or a desk set up in the corridor to work with a student coach who helped me practice addition and subtraction and review the times tables.

The Source for Visual-Spatial Disorders

Britt: The Trouble with Third Grade, *continued*

At home Mom decided that I needed incentives to do better in math; little things like stickers, stuffed animals, movies, and even clothes depended on knowing the multiplication tables. I became more and more frustrated as each incentive was added and then denied, as math goals remained unfulfilled.

With a sinking heart, I wondered what was wrong with me. I kept thinking that I had failed and that I was somehow responsible for not meeting everyone's expectations.

Every week I watched the number of stars grow next to the other students' names.

discuss their fun evening at the picnic.

Spelling was another area of trouble in school. Every week Mr. Palmer chose ten words for the spelling and vocabulary tests. I studied the words as soon as I received them.

To learn the words, I spelled them aloud to myself and then Mom quizzed me. After practicing, I could spell them aloud perfectly. But then Mom made me write them down on paper, and everything went awry. Few of the words, if any, were spelled correctly.

On paper *because* became *becuse* and *surprise* became *surpris*.

> **" I became more and more frustrated as each incentive was added and then denied, as math goals remained unfulfilled. "**

Finally, in May, Mr. Palmer marked the date for the picnic in bold letters on the classroom calendar. Because I was unable to finish the tests, the picnic took place without me.

I sat in class the next day and listened to the stories about who ate the most hot dogs and about the ketchup bottle exploding, leaving a ruined hamburger in its wake. Feeling left out and rejected, I sat quietly with an expressionless face. Trying not to appear upset, I let the other kids

Sometimes I made up a song or poem to help me remember how to spell the words. This technique worked well as long as I repeated and spelled the words aloud, but putting them down on paper remained impossible.

Mr. Palmer would review for the tests by asking students to spell random words from the spelling list and give the meaning. I did relatively well with the oral spelling, missing one now and again, but I never missed a definition.

Although math and spelling eluded me, reading remained exciting and easy. I took pride in reading almost anything in front of me. I also enjoyed having a large vocabulary and surprising adults with the words I could use. Writing was fun when I could dictate a story to Mom, who would transcribe while I told the story.

Later that spring, just before the morning bell, Mr. Palmer joked with the students as he finished setting up his teaching materials for the day on a large counter that sprawled across the back of the room. I listened to the classroom banter, lifting the top of my desk and rummaging for a pencil and notebook. Finding what I needed, I looked around and noticed an unusual mural that engulfed the entire closet door. That's strange I thought to myself, those pictures are new, and I don't remember Mr. Palmer mentioning an art project.

I soon learned that the pictures in question were part of our final math evaluation. At the end of every school year, we had to be tested to see how far our skills in reading and math had progressed.

After visiting the King Tut exhibit in Seattle earlier that year, Mr. Palmer became enamored with the idea of archeology and incorporated a history lesson with the math

The Source for Visual-Spatial Disorders

Britt: The Trouble with Third Grade, continued

evaluation he had planned to give at the end of the year.

"I'm sure that you are all wondering why the front closet is covered with construction paper and tempera paint. Believe it or not, it's your final math assignment," he said. "I want you all to play the part of an archeologist on an expedition, just like Indiana Jones. It's your job to find and to decipher all the hieroglyphs enclosed in the pharaoh's tomb. The painting on the front of the door represents pictures and images similar to those on the walls of Egyptian pyramids."

Mr. Palmer showed us pictures of Egyptian paintings and artifacts as well as further examples of hieroglyphs. "Okay," Mr. Palmer said, "this is how the project works. Everyone will get at least five turns to unravel the mystery message inside the tomb. I've placed math problems for you to solve on 3 x 5 cards across the walls of the closet. Take the answer sheet and a pencil on your expedition and try to figure out as many problems as you can in five minutes. After you have answered the math problems, look at the chart I'm going to give you to decipher the code. It's important to do the problems in order; otherwise the code won't work."

Finishing his lengthy and rather confusing introduction

> *" When it was my turn to go into the closet, I hesitated a moment, then stepped in and the door shut behind me. "*

to the project, Mr. Palmer handed out the answer sheets, the written instructions, and the chart to crack the coded message.

I quickly learned to loathe the closet. When it was my turn to go into the closet, I hesitated a moment, then stepped in and the door shut behind me. I looked up and noticed light emanating from a single bulb in the ceiling. I began searching for the first question by looking around the walls filled with what seemed like hundreds of little cards until I finally found the question labeled *number one*.

Still standing, I attempted solving the problem and recorded my answer on the wrong line of the answer sheet. When I finished, I looked for the second question. Even though I had found the second question while searching for the first question, I didn't remember where I had seen it, so I began the agonizing search process all over again. When my time in the closet was up, I had completed only the first two problems. I had done so poorly, I didn't even try to crack the code.

I escaped the math closet, tears streaming down my cheeks, with a sense of failure and frustration. Difficulty with both completing the math problems and finding the right questions on the wall turned an experience intended to be fun into a nightmare.

ಬಿ ಬಿ ಬಿ

The Source for Visual-Spatial Disorders

Julie: Third-Grade Mother

On that spring day of 1981, I remember I was looking forward to seeing Britt's teacher. Rob Palmer had been a friend for some time. I liked him, and I knew he liked Britt. But then almost everyone liked lively, blond-haired, blue-eyed Britt. She was small for her age, barely showing up on the charts, but she had a deep, strong voice and a big vocabulary that surprised most adults. I knew Britt was having trouble writing her times tables, but when she practiced at home, she could say them perfectly. I just thought the rapid writing of the times tables would come with more practice.

As I had done many times before, I entered the third-grade classroom. Reading books lined the window ledges, multiplication table charts hung above the chalkboard, and spelling words written on red construction paper hung on the bulletin board.

But this day when I sat down across the desk from Mr. Palmer, Rob wasn't smiling.

"I really don't know where to start," he said. "Certain things about Britt's performance just don't make any sense." He said that Britt was in the top reading group, yet she couldn't find a page in a book when directed to do so. "I'll say to the class, 'Turn to page 59,' and every student will be on page 59 except Britt who will be thumbing through her book aimlessly. I usually go over and turn the page to the correct spot, but why can't she do it herself? She might be on page 52 or 129 or 95; there's no pattern to it."

Rob paused for only a second. "She can say her multiplication tables, but when it comes to writing down the answers, she is among the slowest in the class and she makes mistakes." I nodded to show that I had some knowledge of Britt's difficulties, yet I really hadn't comprehended the depth of her problems. He said that she memorized a difficult poem in one night, but had trouble spelling words that were routine for her grade level.

"Now, I don't know what this all means. Maybe she's inattentive because," he hesitated, "well, I understand you and Dave are getting a divorce. Maybe she's distracted." He paused for a second and then rushed on. "Or maybe it's a stage she's going through. I don't know what to tell you, but things just don't make sense."

I thanked him for the conference and told him I would talk to Britt's father about what we might do concerning Britt's problems in school. I smiled as I shook Rob's hand, but I left feeling hurt, angry, and confused. What if it was the divorce? I thought I had done everything I could to keep life stable for Britt and her brother, John. Why hadn't Rob talked more about what a good, smart kid Britt is? I resolved to spend more time helping Britt with math and spelling. And we'd practice finding pages in the book. I'd just have to do more to help Britt deal with this emotionally difficult time and the difficulties with school.

That night I called Britt's father to report what had happened at the conference. He agreed that the report was worrisome. The next day he called Rob, heard essentially the same thing, and decided that there must be something physically wrong with Britt. He thought we should pursue the possibility of a physical problem before we looked into emotional problems. Not trusting Britt's pediatrician, he made an appointment for her to see

> **" I nodded to show that I had some knowledge of Britt's difficulties, yet I really hadn't comprehended the depth of her problems. "**

The Source for Visual-Spatial Disorders   14   Copyright © 2002 LinguiSystems, Inc.

Julie: Third-Grade Mother, continued

another doctor for a complete examination.

It was late spring by the time Britt could get the appointment with the pediatrician across town. The doctor was definite: Britt was a normal, healthy ten-year-old in every respect, except for her size. Yes, she was small, but children came in all shapes and sizes, and he didn't see Britt's shortness as a problem.

Not ready to settle for this answer, her father took her to another pediatrician who agreed with the first diagnosis. School ended. I was happy enough to put doctors and tests aside for a while and take a break from math and spelling drills. So was Britt.

In late June, a third doctor confirmed the diagnosis of the other two, but said there was one more "very sophisticated blood test" that he could do that might reveal something. He didn't say what.

By this time, the kids and I were enjoying the summer—searching for wild blackberries during the day and having campfires with the neighborhood kids in the evening. Britt, John, and a few other kids were building a fort in the woods behind the house and hammers, nails, and boards disappeared into the trees.

Each day I tried to do some academic work with the kids. For Britt, it was practicing multiplication tables. I knew Britt wasn't crazy about math, so I tried to turn it into a game with small rewards for successful

> **" I hung up the phone and stood dazed, feeling both shock and despair. What now? "**

learning of a multiplication table. When she said them orally, Britt knew the multiplication tables quite well, but she still had trouble writing them down. Sometimes the rewards worked; sometimes they didn't. Sometimes she knew 5 x 4 = 20, but puzzled over 4 x 5. So John, who was seven, wouldn't feel left out, I wrote out pages of addition and subtraction problems and John whizzed through them, asking me to please make them harder.

But the schoolwork was only a small part of the day. We read, played cards, and picnicked on the beach at a nearby state park. We were all enjoying sleeping late and spending time outside.

By midsummer the bubble burst. We had the results of the blood test and we had the answer: Turner's Syndrome. Turner's, an accident of conception, is a genetic disorder. Turner girls are missing part or all of an X chromosome. They tend to be short in stature and given to heart, kidney, thyroid, and reproductive problems; as well as prone to learning disabilities. They can never have children, and without treatment, they die young. The doctor who ordered the blood tests recommended that we start with a pediatric endocrinologist immediately and perhaps, eventually, a learning specialist. He would send the results of the blood analysis to Britt's pediatrician. No, there could be no mistake; the diagnosis was positive and final.

When the call came, I had been sitting on the deck, sipping an iced tea, reading a book, and enjoying the warm Northwest afternoon. From my deck, I could see the blue water of Hale's Passage gleaming in the sun, and I could hear the kids in the woods as they worked on their fort. I hung up the phone and stood dazed, feeling both shock and despair. What now? Do I call Britt inside? And if I do, then what? She's not actually sick. Hot tears rolled down my cheeks as I straightened the deck chairs and poured out the tasteless iced tea. I took out the garbage and came back into the family room to straighten pictures. Maybe they made a mistake. How could they know for sure? "Irreversible" and "final" kept floating through my mind.

The Source for Visual-Spatial Disorders

Julie: Third-Grade Mother, continued

Who could I talk to about this? Who would understand something I didn't understand myself? An accident of conception, but what part did I play in that accident? Was it my fault? I had never imagined that Britt had a serious problem. My mother had died four months before Britt was born. I felt like I needed her now.

I decided to call my dad. I wasn't sure what I was going to tell him, but I needed to hear a sympathetic voice. I dialed the number of his beach place on Vashon Island. It was July and I was sure he would be there.

"Hi Dad," I said.

"Oh, Julie, I'm so glad you called. It's a great day here on the island. Why don't you pack up the kids and come out for dinner?"

"Dad, I have some bad news about Britt. You know those tests she had done? Well..."

He interrupted, "She is the cutest little thing, and she does love her grandpa doesn't she?"

"But..."

"So can you come for dinner? There's a 5:30 ferry."

"No, it's Wednesday—Dave's night to have the kids, but I'd love to come," I said. Even if I couldn't talk about Britt's problem, I just wanted the comfort of family.

"No, let's wait until you can bring my little darlings too."

I sat on the floor by the telephone and cried. Even through the divorce, I had never felt this bad. I heard Jason's mother calling and knew Britt and John would be in before long. I took out some cold lemonade and put four cookies on a plate. I had to do something. I just didn't know what to do.

The next morning, I called our pediatrician, Dr. Medlar, who said he had received the results of the blood test and was surprised at the diagnosis. He would be happy to continue to manage Britt's primary care. He recommended a pediatric endocrinologist at the Children's Hospital and would have his nurse call to set up the appointment. After hearing about Britt's unusual problems at school, he suggested we have her examined by a learning specialist. In fact, he knew a woman who had just started a clinic for people with learning disabilities. He recommended that we have Britt tested to find out the nature and extent of the learning disability. I set up an appointment for Britt to see Dr. Medlar in late August after his vacation.

Although I was happy to have answers, I went through the grieving process—denial, anger, depression. At the beginning of the summer, I had a bright, healthy, beautiful child. Now I saw a beautiful little girl who was doomed to a life filled with medical and learning problems that would only become worse as the schoolwork became harder. I cried often.

> **"** *Although I was happy to have answers, I went through the grieving process—denial, anger, depression.* **"**

I also gathered information about Turner's Syndrome in my attempt to understand. Although none of the literature suggested that I had caused the problem, I spent time analyzing my pregnancy. Could it have been that cup of coffee I drank in my second trimester or the wine I drank before I knew I was pregnant? I blamed myself though the doctors told me repeatedly that nothing I did before or after Britt was born caused this condition.

I was depressed and fatigued because now I was not only a single mother, but a single mother with a child with very special needs.

ଓ ଓ ଓ

The Source for Visual-Spatial Disorders

Carol: Long Long Road

The early morning phone call from Dr. Medlar reached me as I finished unpacking the new testing supplies. As he described a little girl he was referring, he interspersed his conversation with assurances that our learning clinic could help her. He briefly described her recently diagnosed medical condition, Turner's Syndrome, and said he would let me know more from the medical literature soon. I had never heard of Turner's Syndrome, but the excitement in his voice conveyed a message. This girl was very special.

None of the books in my cramped office contained any additional information about Turner's Syndrome. That evening I looked in my library at home. By the time I closed the last book, I had begun to wonder if anyone in education knew anything about this—this condition or problem or whatever it was. Could Dr. Medlar be wrong? Not likely. His diagnostic skills were highly respected. Besides, he had referred to specialists at the University Hospital which probably meant a challenging case. Could this really be new territory? Where could I find out more? As I replaced the books, I felt a rush of energy.

For 13 years, I studied learning as a parent, a teacher, and finally as a clinician. Children had given me glimpses of their problems, and teachers had shared their concerns.

My interest began when my son Allen, then in first grade, was diagnosed with a learning disability. We had no indication of any problem prior to our conference with school officials. Of course we knew that even at six, Allen did not speak clearly, but his speech was improving. Since we could understand him, we weren't concerned.

The pronouncements at the conference were such a shock. I thought back to that moment. Shy, left-handed Allen slowly thumbed through books as he waited for us at a small table in the hall. I remembered the silver-haired principal welcoming me to the conference and calmly explaining that our first-grade son was dyslexic and wouldn't learn to read. His teacher nodded her agreement. They described the "symptoms" that led to this diagnosis of delayed brain lateralization: left-handed, unclear speech, avoided reading ... "Dyslexia," they concluded.

I was too stunned to ask questions. Their recommendations were extreme: take him back to the six-month stage and go through all the successive developmental stages. "Force his brain to lateralize," they said. "Exercises. He needs to hold things on his tongue at night ..." My voice was saying, "We will do anything to help Allen."

We were overwhelmed with dismay and despair, the despair of parents who are searching for help, but finding misinformation and even indifference.

At home, I watched Allen and his friend play in the backyard. Was his brain injured during his difficult birth? Maybe that fall when he was eight months old ... Maybe the asthma ... Maybe the new baby had upset him emotionally .... I voiced all my distress to his father, who countered with calm assurances. All my questions were met by,

> " *We were overwhelmed with dismay and despair, the despair of parents who are searching for help, but finding misinformation and even indifference.* "

"We will handle it. He will be okay."

Even after all my work in education—teaching and creating kindergartens for low-income kids—I did not know about dyslexia. In the local university's library, I found one book on the subject, but nothing in that book described Allen.

The Source for Visual-Spatial Disorders

Carol: Long Long Road, *continued*

So, as the principal had recommended, we tried the rubber bands. I sat on the edge of Allen's bed with the packet of tiny elastic circles on my lap. Blue-plaid bedspread tucked under his chin, his brown eyes never left my face as I explained that he was to hold the rubber band against the roof of his mouth with his tongue. We practiced.

The first one "disappeared" when he swallowed as he asked me, "What happens if I eat it as I sleep?" We tried again. In the morning, the tiny band was beside his face on the pillow. Allen rubbed his eyes and kept repeating, "I tried Mommy, I really tried."

I put the packet of bands in the back of a drawer with other useless objects.

Over and over I reviewed the school conference. They saw that Allen spoke unclearly, was left handed, and shied away from reading in front of others. From this, they concluded dyslexia—but the label did not ring true. Moreover, their proposed "solutions" were too stressful for Allen.

In the middle of our search for answers, we moved to Washington State. I studied more about learning problems and confirmed that Allen was not dyslexic. With speech therapy, Allen's words became understandable and his grammar became standard.

I began graduate study searching for answers. Focus on perception plunged me even further into studying about the brain and learning. I got hooked on the unusual learners, those kids who are smart, but do not fit into most classrooms. I often wondered how many other children were, like Allen, being misdiagnosed.

About this time, current newspapers and magazines were filled with articles on split brain research. Words like *right-brain* and *left-brain* entered our vocabulary. Many teachers were interested in this new information on learning, including the revelation that right and left hemispheres functioned differently.

> **❝** *I often wondered how many other children were, like Allen, being misdiagnosed.* **❞**

To respond to their interest, I taught graduate courses for several universities. The classes filled quickly, drawing educators from a considerable distance. Teachers wanted help with the "problem cases" they brought to our classes. "Why couldn't Missy read?" "Why was Matthew's handwriting so delayed?" And on and on.

These questions and an incident in a graduate course I was teaching helped convince me that a learning clinic was needed in this community.

The class, "Teaching Techniques for the Classroom," focused on instructing students who needed non-typical methods. A teacher in one of the classes asked for help. She was concerned. Clayton, a second-grade boy, was not reading or making noticeable progress. By now he had been in the same reading program for nearly three years. Her principal wanted the child to go to summer school in the same reading program and then repeat second grade. We talked about the child. Yes, he was regular in attendance. Yes, he had a normal vocabulary. But he couldn't remember the names of several letters. And he seemed unable to learn the sounds that went with letters. His reading program was dependent upon the one thing he couldn't do—remember sounds. "But," I said, "some children do not learn phonetically, so how can he recognize the sounds, retain them, and later assemble them into words?" The answer was that he couldn't. He needed a different program.

Where could Clayton turn? Where could all the Allens and Claytons turn? They needed someone to look carefully at them as individual learners and

The Source for Visual-Spatial Disorders

to work with them until they succeeded. They needed an alternative.

So after trying many steps—personal research, working in several school systems, graduate study, and problem solving classes—I joined others to start a clinic that would serve children and adults with learning problems.

Physicians, teachers, and psychologists helped with the groundwork of establishing a philosophy and developing criteria for assessment and instruction. Next we worked with children at no cost to refine our methods. Finally, we taught an intense and lengthy 11-credit workshop for specialists in medicine and education. These specialists critiqued all of our ideas.

Only after all of these steps did we open the clinic. Our approach was simple: discover each client's strengths as well as obstacles; describe how that individual needed to be instructed; teach him or her; and work with the educators and medical people involved.

The clinic had only been open for four weeks, but several children and adults had already begun work. The modest office included a tiny waiting room set off by a bookshelf and a small workspace that held two desks, a table, and several files.

> **❝** *I thought about the intriguing challenge in the call from Dr. Medlar as I opened the file drawer, took out a new folder, and carefully printed 'Turner's Syndrome' on the tab.* **❞**

I thought about the intriguing challenge in the call from Dr. Medlar as I opened the file drawer, took out a new folder, and carefully printed "Turner's Syndrome" on the tab. What a long journey this has been. I smiled in anticipation. The next stage of the journey could be very interesting—very interesting indeed.

ಬ ಬ ಬ

Julie: The Beginning—Dr. Medlar's Office

Allenmore Medical Center is one of those huge medical complexes—a conglomeration of doctor's offices, hospital, outpatient services, and medical laboratories. Brown and ugly in the late August rain, it sprawls over several city blocks.

Inside, Dr. Medlar's office was wedged between eye surgery and obstetrics. Here, we waited for Britt's appointment with Dr. Medlar—the first appointment since the Turner's diagnosis. Britt walked across the room to look at a baby swaddled in pink, sucking on a bottle. Returning to the seat next to me, she hopped up and whispered, "That baby is really cute. It's a girl." Britt picked up a *Parents* magazine, her feet sticking straight out from the chair, her back straight against it. She started reading an article on baby nutrition, stopping often to ask me for details about what I had fed her when she was a baby.

"Did I eat cereal before I ate pears?" she asked. "Did you breast feed me?" "Why didn't you breast feed me? The magazine says it's healthier for babies to have breast milk."

By this time, all of the adult eyes in the room were on Britt. She was the size of the four-year-old boy next to her, but her language and her reading ability were an indication that she was much older. The boy with the runny nose brought Britt a plastic toy truck. She lifted her head from her reading, looked at him scornfully, and went back to her reading.

The nurse opened the door to the examining room area and called, "Britt Neff." Britt jumped off her chair and hurried toward the door to the examining rooms for her turn with the doctor, and I followed behind. The nurse led us to one of the rooms and explained that they wanted to find out why Britt might be having difficulties in school.

She handed Britt a stack of half sheets of white typing paper, a pencil, and a pile of black and white drawings of various kinds of objects. Some of them were basic shapes; others were stereotypic pictures of houses, flowers, and common objects. She asked Britt to draw the pictures as best she could on the white paper and said she would return in a few minutes. Britt took a seat at the table, and I sat in the corner and opened a book of essays. When I looked up, I saw that Britt was tracing the pictures, quickly but carefully following the lines to reproduce as exactly as she could the pictures that had been given to her. I wondered about the tracing, but I hadn't paid close attention to the directions Britt had been given. "Oh Britt," I said, "I don't think the doctor wanted you to trace them."

"He wants them right, doesn't he?" Britt replied. "This is the only way I can get them right."

"I guess I really don't know what the goal is. Anyway I have to go to an English department meeting. I think your dad is in the waiting room. He's going to talk to the doctor and then you'll pick up John at Grandpa's house and go to dinner." I gave Britt a kiss and hug and assured her I would be waiting for her when she came home after dinner.

I left the medical center feeling relief and apprehension. This was the first step in figuring out the full extent of Britt's learning problems.

> *She was the size of the four-year-old boy next to her, but her language and her reading ability were an indication that she was much older.*

ಌ ಌ ಌ

Carol: Meeting Britt

Catching my breath after running up two flights, I opened the back door of Dr. Medlar's office and slipped inside. He had insisted that I come right away "to meet the young girl with Turner's Syndrome." Unobserved, I peered through the half-open doorway to his busy office.

A blond head bobbed behind the glass of an examining room. A small girl was seated at the table, pencil in one hand, paper pinned to the table with the other. She drew rapidly, thrust the page at the nurse who was watching, and reached for another sheet of paper. Her moving lips testified to continuing chatter. Animated, head cocked to one side, eyes punctuating each action, the little girl radiated charm and confidence.

Dr. Medlar was in his office talking to a blond man in the straight-backed chair opposite him. He waved me inside, made introductions and said, "We need some help here. We need to figure out how to help with these school problems."

Minutes passed as I listened to Dr. Medlar explain a medical condition that he found interesting. "Unusual . . . first in his practice . . . health pieces manageable . . . talked with the research fellows at the U. . . ."

The receptionist walked by the office doorway for the third time. "Okay, I'm coming," the doctor called over his shoulder as he backed toward the doorway. "Now Carol, I want you to tell Britt's dad what you do. Her problems right now are with school. I told him you could fix that." The door closed in his wake.

I broke the silence that had settled in the room. "Your daughter is failing at school?" I asked.

> **" *She bounced across to meet us, hand out, waiting for mine.* "**

He shifted his eyes from the tree outside the window to the ceiling and cleared his throat. "Well, not failing. At least, I don't think so. But the doctor says she needs help. Maybe you need to talk to Britt's mother. We're divorced. She knows everything about Britt and school."

I gave him my card and asked him to pass it on to Britt's mother. We rose and moved toward the door.

Perched on a chair in the waiting room, the blond child I had seen before was reading a Dr. Seuss book to two boys.

"Just a minute, Dad, I want to finish reading this page," said the girl as her words tumbled out and over her rapt audience.

"Britt, I want you to meet this lady. She's a teacher."

Britt slipped forward off the chair until her feet touched the floor. She bounced across to meet us, hand out, waiting for mine. "Hi, I'm Britt. I have two cats, Muffin and Blackie. Do you have cats? I'm reading *The Happy Golden Years.* That's the last book in the Laura Ingalls Wilder series. Do you like to read?" The words tumbled out as her small hand gripped mine.

I looked at her lively face and thought, "What kind of problems could this spirited, assertive child who 'loved to read' possibly have? Could this be a mistake?" Of course, even gifted kids can have learning problems.

Her bright smile stayed with me as I returned to my office to clear my desk. I wondered if or when I would hear from Britt's mother.

ಙ ಙ ಙ

Julie: Meeting Carol

About 8:00 in the evening, I heard a car pull up in the gravel driveway. David entered through the back door with Britt and John and then handed me Carol Stockdale's business card.

"How was the meeting?" I asked. "What did she say? Did she think she could help Britt?"

"She wasn't sure. She'd like to talk to you. I told her you'd call her when you had a chance," David replied. He turned to the children, evidently anxious to be on his way. "Bye Britt, bye John. See you next week," David said, hugging the kids good-bye.

I put Carol's card by the phone and figured I would call her tomorrow. With the new school year only a week away, I was nervous about what fourth grade would bring for Britt.

The next morning the phone rang before 7:30. It was Dr. Medlar phoning to tell me about the meeting with Carol that I had missed. "You need to call her right away to see about getting some help for Britt's learning problems. This problem isn't going to go away." I told him I had the number and thanked him for the call. I was surprised at the urgency in his voice. I had never before received a call like this from a doctor.

Figuring that Carol Stockdale would not be in her office so early, I waited until 8:00, filled my coffee cup, and dialed the number on the card.

"Oh, yes, Mrs. Neff. I met your daughter and your husband, ah, former husband, last night. What a beautiful child, and so bright," Carol said with a smile in her voice.

"Thank you. She is bright, but she also had a hard time in school last year," I said. "Dr. Medlar said you might be able to help her. I don't know what to do about her problems with math and spelling."

"Well, we can try to help her, but I'm not sure yet what we'll be able to do. Perhaps you and I could meet fairly soon to discuss Britt's learning profile and your options for getting Britt some help." Carol's voice was easy and mellow, competent, but not overly confident. I liked her immediately. Best of all, she offered hope for Britt.

We set the appointment for that afternoon. I was happy to get started and pleased that I didn't have to wait a week or more for the appointment. I dropped an English muffin into the toaster, feeling a great sense of relief.

Later that morning, Britt, John, and I took buckets and headed down the road to pick blackberries. Along the road, the vines formed a tangle of green, and the berries hung heavy on the bushes in large clumps. In a short time, we had filled all three containers.

"Let's go to Marianne's house. She might like to see our blackberries. You could have a cup of coffee with her," Britt suggested. John and I agreed that this was a good idea, so we continued down the shoulder of the road in the hazy August sunshine. Soon we were knocking on Marianne's front door, displaying our full pails of berries.

While Britt and John joined Marianne's children in the playhouse, Marianne and I sat on her deck, and I shared the conversations I had had with both Dr. Medlar and Carol Stockdale. Marianne and I were long-time friends, and I had told her my concerns about Britt's problems in school and the diagnosis of Turner's Syndrome.

"You know, Julie," Marianne said, "we love Britt, but you have to realize that her condition is permanent. I know this

> **"** Carol's voice was easy and mellow, competent, but not overly confident. **"**

The Source for Visual-Spatial Disorders

Julie: Meeting Carol, *continued*

is hard, but you may have to lower your expectations."

I felt the knots in my stomach. "I may have to do that, but for now, I can't give up. I'm really not willing to accept that she's doomed to fail until I've done everything I can," I said.

"That might be expensive," Marianne replied.

"I know," I said, "but we'll just have to manage."

I changed the subject to plans for Labor Day, but I felt the knot tighten. In the back of my mind, I knew the reality of learning disorders, but I needed someone to share the good news of finding Carol, not to be reminded of the possible abyss.

I finished my coffee and called Britt and John, but as we headed home, the joy I had felt while we were picking blackberries was gone. I began to realize that even Carol was not going to have simple or easy answers for us.

Although I was not as optimistic as I had been, I was hopeful about my first meeting with Carol. I arrived at her office at 3:00. Carol was a tall, sturdy woman with neatly trimmed blond hair and a manner that was warm, friendly, and optimistic. But as she shook my hand, I had the feeling that she also knew what she was talking about. In fact, she was the first person I had

> **" *She seemed as interested in what Britt could do as she was in what Britt couldn't do.* "**

met in two years who seemed to have any idea about what Britt had been going through.

We talked for almost an hour. I related the trauma with third grade, Britt's decreasing sense of confidence with school, and, of course, the recent diagnosis. Carol asked questions that suggested she had had significant experience with learning problems and that she was genuinely interested in Britt's problem. She wanted to know about Britt's early childhood. I was happy to tell her about how Britt played, what toys she liked, and how her language ability developed early. She seemed as interested in what Britt could do as she was in what Britt couldn't do. Her curiosity about Britt's strengths was a sharp contrast to Britt's teacher's concern about her weaknesses.

Carol said the first step would be diagnostic testing, but even before that, she'd like to get acquainted with Britt. Then we would discuss a plan for testing.

Finally I asked, "How much will the assessment cost?" Almost apologetically, Carol said that it would be about $500.00. She must have seen my face fall because she quickly added, "Of course, we can work out some kind of payment plan. Maybe your insurance will cover it?"

I didn't know what I would do about the $500. For a single mother with two children, it was an enormous amount of money, but I was certain that something had to be done to help Britt with school. I would not let her fail no matter how much it cost.

I set the appointment for Britt to begin working with Carol.

ಙ ಙ ಙ

Carol: Meeting Julie

I was startled when the phone rang shortly after my early morning arrival at the clinic. Dr. Medlar's voice on the phone reflected his excitement as he spoke quickly about the girl I had met yesterday at his office—the little blond girl. I looked at the notepad and sorted out some of his phrases that I had written down in an attempt to anchor the conversation. "Cell something," "medically manageable," and underlined three times as he circled back to the point, "*absolutely permanent medical condition.*" When I broke in with questions and doubts about my role in a "medical condition," the doctor paused. He chuckled. "Remember she has learning problems and no one seems to know what to do about them. You might want to see the drawings she did for me. Another thing, you need to talk to her teacher. The school thing isn't working at all." I could hear another phone ringing in the background. He closed with, "Gotta run, Carol. Keep me posted."

I put the phone back on its cradle, straightening the cord and lining it up with the edge of the desk. "Lord help me," I said as doubts swept over me. Could I understand this girl's problems, much less help her?

Just before 3:00 p.m., I heard the clinic's door swing open. Mrs. Neff was prompt. Picking up the clipboard of standard intake forms from my desk, I moved around the divider to meet her in the tiny waiting area. Briskly moving a few steps into the room, the woman returned my greeting and identified herself as Julie Neff. She accepted the forms, slipped into a straight-backed chair, and eased her bulging briefcase to the floor. Smoothing the gray wool of her skirt, she evened the stack of papers on her lap, removed the pen from the top of the clipboard, and began filling out the form that asked for billing information and her daughter's history. Her pen moved rapidly across the page.

Paperwork complete, Julie moved to the chair opposite me.

Her questions were specific; mine, up to this point, had been general. Soon she knew that I had taught in the public schools, had sons who triggered my interest in learning problems, had spent years digging for greater understanding of learning, of the brain, and of recent "breakthroughs" from split brain research. I could feel her making careful mental notes of each response even though her notebook remained closed. Part of me stood back observing the mutual interview. Curly hair. Understated professional clothes. Well connected. Lawyer maybe. Not an accountant. Not sales. She certainly appeared self-contained. Notice how she paused to gather her words, eyes focused either on the beige carpet or the corner of the room above my head as she formed her questions, but she looked right at me when she asked them. Could be an academic. Her hands rested calmly in her lap. Small hands. Gradually I realized she was small. No wonder Britt was so little. Head tilted to the left, Julie leaned into listening as if she heard more than I was saying.

> **"** *I could feel her making careful mental notes of each response even though her notebook remained closed.* **"**

I asked about school.

She glanced to the floor after several seconds of focus above my head. <u>I pushed away my tendency to fill the space with words.</u> She then described Britt's fluctuating achievement. Last school year, Britt apparently was at the top and bottom of her third-grade class.

Julie spoke slowly and seemed to be measuring her words. "Britt loves to read—especially out loud. In fact, we read to each other. She really likes that. She likes to talk about the stories and the people. Britt

Carol: Meeting Julie, *continued*

adds ideas to what could happen and understands even difficult narrative." Julie paused, "I think she 'moved into' the *Little House on the Prairie* with Laura—the whole thing was so real to her. Britt would take on the role when she was playing by herself at home. I know she understood theme and motivation better than some of my students—I teach literature and writing at the University.

"Yet Mr. Palmer, her teacher last year, said Britt was doing poor work in several areas. Her spelling is inaccurate. Often her work is partially done or incorrect. And she is having a terrible time with math. She still doesn't know her times tables."

I was still stuck on *Little House On the Prairie* as Julie continued to talk. Britt was reading such a book in the third grade! How precocious. If this were true, she certainly had reading skills beyond most third graders. My mind finally gathered the rest of her words. Problems with spelling and math. Really? I wondered why. Julie was describing other concerns in a matter-of-fact tone of voice. In contrast to her voice, her hand gripped the wooden arm of the chair.

Julie described the current emotional effects of Britt's school performance. "Stomach aches, headaches.... Happy, buoyant little girl was now often quiet and pensive. She was anxious about spelling and math."

I looked down at my notes and added an underline to the phrase, "uneven performance in the classroom." I looked at the words and wondered if there was an easy explanation for all this. Like an incompetent teacher or maybe an indulged child who only did what she wanted to do.

But the indulged bit did not jibe with the excessive worry pattern the child was demonstrating over math and spelling.

> " *I looked at the words and wondered if there was an easy explanation for all this.* "

More history was surely needed before we took one more step along this road.

I said, "Julie, tell me about Britt's early life. What was she like during infancy and preschool? May I see the forms you completed?"

As Julie handed the papers across the desk, her face softened into a smile as she recalled Britt's infancy. She said, "Britt was our first, a full-term baby, no problems that really were a concern. She was so alert—big, blue eyes, blond curls. It was difficult to get through the aisles of the grocery store because everyone commented about her. Always rather small. She was a little late to crawl and walk, although she certainly was within the normal range. But her language was clear *and* early."

I looked down at the history on my desk. "She talked at eight months! You don't mean it!" I exclaimed. "She said words at eight months and spoke in sentences at a year? I've never heard of such a thing!"

Julie smiled as she responded, "Yes, Britt was sitting in her infant seat when she said her first words." She added, "She was small for her age so no one, including me, could believe the language coming from her."

Julie continued, "When Britt was about ten months old, before she was walking, her grandmother had a plastic toy with different shaped pieces. Britt learned the names of the shapes—pentagon, trapezoid, octagon—and she could say them, put them into sentences, and manipulate the toy. Her grandparents loved to have her perform for them."

Julie thought of another event. "When Britt was five, a friend told me she loved to have Britt come to play because she was so good at organizing the other

The Source for Visual-Spatial Disorders

Carol: Meeting Julie, *continued*

kids into make-believe games. 'If Britt is here,' my friend said, 'the family room is the farm; the living room, the town; and the front yard, the prairie.' She never had to worry about entertaining the kids when Britt was around."

"Oh," I questioned, "tell me about her friends. Does she have many?"

Julie replied, "Yes, she has several friends in the neighborhood and friends from our old neighborhood. She is often the leader in their play."

I made notes and looked up as Julie added, "Because she's always been so bright, it's really hard for me to believe that she's now having trouble in school."

Assured that the early language examples were verifiable dates and that Britt's language still was unusually strong, I circled the question mark I had made and wrote, "Check this out" in the margin.

At the end of our discussion, Julie knew that I offered neither pat answers nor simple solutions. I knew that Britt was a dramatic, imaginative child who loved to talk, read, and play "let's pretend," but who despised math, disliked spelling, and avoided drawing

> **❝** At the end of our discussion, Julie knew that I offered neither pat answers nor simple solutions. **❞**

and puzzles. Oh yes, and she was worried about school—especially about math.

Soon Britt and I would really get acquainted through the diagnostic workup. What a terrific opportunity! Never, ever had I heard of anyone like this. What was Britt's learning pattern? Always, there was a pattern. Well, anyway, always before. Language was usually an indicator of how a child would learn. Apparently Britt had exceptional language. How could she be having such problems?

I caught myself staring at my pencil drawing circles all down the margin of my page of notes. Minutes had passed. "Britt," I said to the empty space recently occupied by her mother, "I am going to find out what makes you tick."

ಬ ಬ ಬ

Britt: Struggling to Understand

Fourth grade remains a blur as I have tried to block everything from that year out of my mind in order to continue enjoying school and learning. Sometimes the line that separates constructive criticism from opinions of you as an individual becomes obscured. Words meant to describe performance eventually end up characterizing the person, and the label sticks.

When you're a kid, the most important thing in the world is to fit in with the crowd and the only thing worse than being picked on by other kids at school is being singled out by a teacher. Even worse is not knowing why you are being treated differently. Subtle glances and innuendo, whether conscious or unconscious, cause panic and anxiety for any poor student merely trying to blend in with the crowd at school.

I knew that I was not fully living up to the teacher's expectations even though I tried desperately to complete all the assignments perfectly. Having a sense of self is very important, especially when teachers criticize. It is easy to take their comments and impose them on your own view of yourself. If you know that what they say is not true but coming from ignorance, it becomes easier to see the full picture. I told myself over and over that I was trying and that next time I'd get it right.

My brother John and I made our way to the bus stop at the end of our neighborhood by alternately walking, running, and pushing each other the entire three blocks. Since this was the first day of school, we were excited to see our friends and new classrooms. Going back to school always held the promise of a fresh start: new friends, new books, new teachers, and best of all, new school supplies. Fourth grade was no exception. At the bus stop, we greeted our friends and talked about our summers, swapping any information we had about our teachers. "Oh yeah, I had Mrs. Owen last year for fourth grade." One of the fifth-grade kids at the bus stop offered. "She's really strict, but as long as you follow directions, she's okay. She gives out a lot of homework."

"More than third grade?" I asked. "I had Mr. Palmer last year and he gave us plenty of assignments to take home," I explained.

"Just wait and see for yourself," the fifth grader warned and turned around to grab his backpack and sack lunch off the bench. I put the predictions out of my mind and gathered my lunch and notebooks as a rumble signaled the bus's approach.

Once the bus arrived, we gathered into a small group, shoved our way onto the nearly empty bus, and found our respective seats. Forty-five minutes later, we arrived at school, the bus now packed with noisy elementary kids screaming and shouting to each other at the top of their lungs.

Even though I had been at my elementary school almost three years, I was still worried that I'd be unable to find my way to the new classroom as it was in another part of the building. Anxious to find the right room, I hurried down the hallway trying to keep my notebooks, paper, pens, and pencils from falling out of my arms and spilling onto the floor.

"Hmm," I thought to myself as I made my way down the unfamiliar hallway. "I think this must be the room." Then, finally looking at the door, I noticed the name in huge letters across the top of the door, "Mrs. Owen." "Okay, this is it." Taking a deep breath, I opened the door.

Hoping to find friends among the sea of faces gathered in the classroom, I paused in the doorway to look around before sitting down. Recognizing only a few students, none of

The Source for Visual-Spatial Disorders

Britt: Struggling to Understand, continued

whom I liked very much, I found the closest empty seat. My heart and mind were full of enthusiasm and high expectations. Noise from desks and chairs sliding across the tile floor mixed with excited chatter and shouts from students as we settled into our fourth-grade year.

Mrs. Owen called roll, efficiently moving through her list provided by the front office. "Now that I have taken roll, I want to arrange everyone alphabetically. Find your name and then find your desk according to this seating chart that I've drawn," Mrs. Owen said, pointing left to a small bulletin board next to the door.

As the students shuffled around the room searching for their new places, I was lost. Standing to the side, I planned to let everyone else find places, and then I would fill in the empty desk. "Oh, no," I thought to myself as I saw Mrs. Owen heading over to me. "Why aren't you at your desk?" Mrs. Owen demanded.

"I, umm, I'm still looking for the desk. I think it's in the third row?" I said with a concerned look on my face.

"No, it's not. It's in the second row," she said, disgusted with my answer. "It's the third from the door," she instructed, pointing to her left and then turning abruptly to make her way back to the front of the room, shaking her head and grumbling something under her breath. Embarrassed and ashamed, I made my way to my seat and quietly slipped the pens, pencils, and notebooks into my desk.

With everyone seated in order, Mrs. Owen felt safe to proceed with her instructions. She gave out orders and made her expectations extremely clear.

"All you have to do is copy the assignments off the board and complete the assigned work. I anticipate that all of you will learn quickly how to handle these assignments on your own. Also, I will post the lesson schedule for each day on the bulletin board next to the chalkboard," Mrs. Owen said with a serious, no-nonsense look on her face.

While handing out the textbooks we would need for the year, she described all of the subjects we'd be covering. Of course there was reading and spelling, art, music, and math. But, most important of all, this year we were going to learn about science.

"These science books that I'm giving you may seem a bit complicated, but we are also going to try experiments both inside and outside the classroom in order to better understand the world around us," Mrs. Owen proudly announced. "Put these books in your desk and we will look at them later."

"Now, for the first lesson, I want to start off with some spelling and vocabulary words. Write them down and we'll have a short quiz next week," Mrs. Owen said as she wrote on the board the words that she wanted us to learn.

For every one word I was able to copy correctly, Mrs. Owen had already written five more words on the board. After she had written 20 vocabulary words on the board, she began to erase them and move on to the next word list.

"Wait!" I gasped as Mrs. Owen picked up the eraser and began to wipe away the assignment. "I haven't finished copying that list yet," I said with an anxious tremor in my voice.

She shot me a scornful look, implying that I had been goofing around instead of paying attention. "All right, I'll leave them up on the board for now. But don't expect any favors next time," she said as she began to write the next set

> **" She shot me a scornful look, implying that I had been goofing around instead of paying attention. "**

The Source for Visual-Spatial Disorders

of spelling words on the other side of the board. While I was copying the entire vocabulary assignment unevenly into my folder, Mrs. Owen excused the class for lunch and recess.

Filing into the room after recess, which I had spent walking through the wooded area of the playground, we settled back into our seats. Once we were all quiet, Mrs. Owen started the next lesson.

At three o'clock, the first day of school ended and I let out a sigh of relief. Now I would be able to head home and try to forget that my teacher seemed to despise me. I raced to the coat closet as fast as I dared to gather my belongings and catch the bus. Fighting my way through the narrow aisle on the bus crowded with elbows, arms, and backpacks, I found a seat near a window. I sat staring out the window, hoping that no one would bother me. Hearing one of the kids behind me yell "Hey! Short stuff! Shrimp!", the hair on the back of my neck began to bristle. I let the comment slide and ignored the rest of the jostling and crude jokes coming from the older kids in the back seat. As the bus rolled to a stop near the end of the bus route, I gathered my school bag and coat, grateful for the escape into the lingering warm afternoon sunshine.

The first spelling test of the year was, of course, horrible. Even after all of the practicing I had done that summer, and even after being quizzed at home before the test, I still managed to make lots of seemingly "inattentive" mistakes.

My mistakes were interpreted by the teacher as careless and intentional. She decided that I was a lazy student who refused to pay attention and asked too many questions about the schoolwork.

Fall progressed and the inevitable change in weather marked the passage of time. As the days turned short and gray, so did my attitude toward

> **" *My mistakes were interpreted by the teacher as careless and intentional.* "**

school. Just as fast as the leaves began falling off the maple and elm trees that lined the playground, assignments became harder for me to complete. Crumpled notebook paper littered my desk and returned assignments with red, green, and orange ink were stuffed inside my desk to be swept away later. Unrelenting criticism flowed from Mrs. Owen as I tried and failed to meet her requirements. Sitting at my desk, I diligently tried to follow Mrs. Owen's instructions as she scribbled reading, math, and other assignments on the board. I raced to copy them from the board onto my own paper, continuously making mistakes as I went along and hoping that I would finish before she moved on and started to erase the information. Fear set in as I struggled to transfer the information from the board to the ruled page on the desk in front of me.

None of the work I did in that class was ever right. Each time I copied stuff from the board, I had to start all over again, erasing the word or number three or four times before moving on to the next part of the assignment. By that time, the paper had become thin and worn from the eraser marks. As I erased and started over, I fell further and further behind, ensuring that I would be rebuked in front of the class for my performance. Already feeling as if I had failed, Mrs. Owen compounded the situation by pointing out to the class that I had not finished the assignment and was holding everyone else back.

"Now, after you are finished copying the words off the board, I want you to look up the definitions of the words and use them in sentences." Mrs. Owen's instructions went unnoticed as I scrambled to finish the next set of spelling words. "Are you still copying the words off of the board?" Mrs. Owen asked as she looked down at my paper with her

Britt: Struggling to Understand, continued

brow furrowed and a frown across her face in obvious disapproval.

"Yes, but I'm almost done now," I said as I copied the last two words. Most of the words on the page were illegible from the numerous eraser marks covering the paper, indicating several attempts to copy the words correctly. Other words contained missing and/or transposed letters. For the first time, I began to doubt myself. Obviously, Mrs. Owen was not pleased with me. Unsure of what I had done wrong, I felt a large lump start to form in the depths of my stomach.

At the end of the day, John and I raced home, lobbing our books onto the counter as we stumbled through the kitchen door. We each grabbed a handful of chocolate chip cookies from the cat-shaped cookie jar perched on the corner of the counter and a glass of icy cold milk from the gallon jug in the refrigerator. We had just finished our snack when Mom walked in the door.

"Hi!" Mom said as she came into the kitchen. "How was school?" she asked as she started to pull things out of the cupboard for dinner.

"Awful," I replied. "I don't like my teacher anymore. I don't think she's very nice," I said, crossing my arms and blinking to keep back tears.

"Now calm down, Britt. It can't be that bad," Mom said with a sigh. "Besides, it will get

> **" Unsure of what I had done wrong, I felt a large lump start to form in the depths of my stomach. "**

better. All you need is more practice."

"But she thinks I'm making mistakes on purpose! I guess she thinks I'm dumb. I'm trying so hard, but I just can't keep up with the assignments on the board." I tried to justify my point of view. Tears were now streaming down my face.

"I don't like school anymore!" I yelled. Feeling a strange mixture of embarrassment, worry, frustration, and somehow feeling that this was my fault, I ran up to my bedroom and slammed the door.

☙ ☙ ☙

Carol: Why Are Easy Things So Hard?

Britt chattered as she bounced into my office for our first appointment. "The toad's named Elmo. Cindy named him. If you don't see him in the terrarium when you come to my school, it's probably because he burrowed into the mud and moss."

Britt was filling me in on the terrarium table in her classroom. She chose the smaller chair, pulled it closer to the table, and sat with her left hand gripping the chair seat. She traced vague circles on the table with the forefinger of her right hand. Today we were beginning assessment and I was determined that we would both enjoy the process. The idea was to do enough tests and activities to discover how Britt learned, what she could do, and what she could not do. This medical condition that had so intrigued Dr. Medlar wasn't going to disappear, so we better figure out what if any effect it had on school. The anxious blue eyes followed my every move as I cleared working space by moving a stack of test materials to the shelves beside the table. Now we were ready to start. The first activity needed to be unthreatening.

"It's called 'figure-ground'," I explained, "because some pictures are hidden in the background. Your job is to find the hidden pictures and draw around them." I handed Britt a felt-tip pen as I slid a page that appeared to be drawings of a field in front of her. I knew many fish—8 or 9 at least—were disguised in the pattern.

Britt smiled, "A treasure hunt!" and eagerly started looking for hidden pictures. Several minutes passed as she moved her finger over one after another of the fish, oblivious to their presence in the design. "Um, I think there's a dragon hiding here or maybe a monster," she said as she circled a section of picture that looked quite unlike anything.

"That's all," she said, laying down her pen after one more check of the sheet. Many big and little fish, invisible to her, stared back at me from the test page.

I took the pen and said, "Hey, this old salmon just smiled at me," as I outlined the fish body. She stared at it in surprise, "I guess one fish is there with the dragon," she said. The other fish remained hidden to her.

"Well, Carol," I thought to myself, "figure-ground was a crummy choice to start with. We need something positive."

Britt was telling me a story while I tucked away the page we had worked on and pulled out the green booklet that measures "Visual-Motor-Integration." The events in her story centered on a cat that caught mice and brought them into the house to show off to Britt and her brother.

Her voice reflected the excitement she must have felt when she saw the mouse under the dining room table. "Actually Muffin must think we all adore mice. She kept licking her paw and flaunting that mouse like this was some sort of superb accomplishment."

I smiled back while my mind was cataloging her language. "Complex sentence structure, unusual vocabulary, 'superb accomplishment' is not fourth-grade language," I scribbled on my notepad.

"This test," I explained, "asks you to copy each design in the box under it." Britt's eyes were on me, her head cocked to one side, pencil in hand, as I opened the booklet on the table in front

The Source for Visual-Spatial Disorders

Carol: Why Are Easy Things So Hard?, continued

of her. Three designs were on the page, each with a blank square underneath. The first item was a vertical line. Britt copied the line in the blank box under the item. She quickly copied the second item—a horizontal line. "I bet they get real hard," she commented with a dubious look toward me, as she drew the third item—a circle. Her line wavered, but the drawings were adequate copies. I turned the page to the next three designs for her to copy.

Britt concentrated on the test. The next design was a cross. Britt studied the design and traced it with her finger before drawing it. Next was a diagonal which she drew as a shaky vertical. What a struggle!

A square was easier. Britt announced, "A square has four corners," as she drew them.

As she labored through the pages, copying arrows, diamonds, and overlapping circles, no design was actually easy. Britt struggled to make lines intersect or for a circle to touch a square or for arrow tips to touch the shaft of the arrow. Mercifully, the test was short. I tabulated her score as equivalent to a student about three years younger than she was! Why was copying these simple designs so awful for her? Her shoulders slumped. She knew.

"Well, Britt, at least we got that out of the way," I offered. "Now let's choose an activity that is more fun." Britt asked, "Can we read? Reading is fun."

Britt chose a passage about sailing ships from the alternatives spread out before her. She read it quickly, answered all the questions with ease, and added her own comments. "Everybody knows what a mainsail is. My dad has a sailboat and my mom sails too." The selection she had chosen was from a sixth-grade literature book.

We had time for two more activities. I needed reading levels anyway so I chose a standardized evaluation.

Britt smiled at the clunk when the large Woodcock Test of Reading Mastery was opened before her. She read the vocabulary lists with ease and only paused to stretch when I turned to the comprehension sections. Her fingers tapped the table as she gazed out the window. "I wish we could just read all day at school and not do math.

> **"** Britt struggled to make lines intersect or for a circle to touch a square or for arrow tips to touch the shaft of the arrow. **"**

Where my mom teaches, some classes are just reading and talking. That would be great," she added wistfully.

Britt moved smoothly through other sections of the test, adding editorial comments such as, "'Short.' I hate that word. They should just say, 'not tall.'"

As we finished, I urged Britt to stretch and get a drink of water before starting our final task. Translating the numbers into grade scores indicated that Britt scored over two years above her grade level on the reading tests. She certainly could read well. Now what would be the best measure to conclude our hour? I mulled over the factors. I certainly needed more information, but I wanted Britt to feel encouraged by the testing process. No more drawing today. We needed to know more about visual perception, so a pattern-matching test should be safe and useful. I took the Raven Test of Colored Matrices from the shelf and opened to the first page as Britt stepped back into the room.

"Wish I could be here every day and not go to school," she said as she slid into her seat. I returned her smile as I placed the sheet of patterns in front of her. The Raven looked like a piece of linoleum with a chunk cut out of the design. Britt's task was to look at the six

The Source for Visual-Spatial Disorders

Carol: Why Are Easy Things So Hard?, continued

options at the bottom of the page and choose the piece that filled the hole and completed the pattern. The first items were very easy and she rapidly pointed at one of the alternatives. After the first four items, she began slowing her response. Now she was touching each of the six choices and waving her hand back and forth between them before selecting.

"Wow, this one certainly has a lot of pages," she said fingering the test notebook.

By the time she reached the section that included sequences, Britt had slowed to a snail's pace. Trying to figure out whether lines and arrows went up and down or sideways made her head swim. Her hand movements grew more and more erratic as she searched for each correct answer.

Eventually she reached the final choice and closed the book with a thump. "They worked overtime to make that confusing," she said as she leaned back in her chair.

Britt sighed as I picked up the test booklet and announced that it was time to go. I assured her that she had, indeed, given the test the "old college try" and that almost no one got all of them correct.

She looked up at me through a long fringe of lashes and said, "Well, I'm in good company then," as she pushed back her chair.

"See you soon," in her usual chipper voice floated back through the door as she left.

Britt had missed 44 percent of the items on a pattern-matching test. I had expected her to miss only two or three percent!

I thought over the hour as the minutes ticked away. "What was going on?" Britt spoke mature and precise English. She read as if she were in

> " *She was way ahead and way behind all at the same time.* "

middle school, but she drew and matched patterns like a preschool child. This was crazy. She was way ahead and way behind all at the same time. What a challenge! No wonder she would rather read all day than go to school. That was the only thing that made perfectly good sense in the entire hour.

ಬ ಬ ಬ

Julie: Fourth Grade

October arrived that year with cool nights and warm, sunny afternoons. On one of these afternoons, I arrived home as I usually did just after Britt and John had let themselves into the house. Sitting at the counter in the kitchen, Britt was already spreading peanut butter on saltine crackers, a gallon of milk beside her on the counter.

"Hi, kids," I said, "What's up?"

"Oh, Mom," Britt said, jumping off the stool, "you have to sign up for a parent conference with Mrs. Owen." Britt took a conference form from her book bag and put it in my hand.

"Oh," I groaned, looking at the form. "Every day that week you're out at noon." Early dismissal is always hard on working parents and my kids understood the problem.

"We can go to Grandpa's," Britt offered. "He likes to pick us up and take us out to lunch. I think it's just an excuse for him to eat at McDonald's. Which day do you want to see Ms. Owen? So what's for dinner?" Britt paused only momentarily to spread more peanut butter. "Why don't you go on Tuesday afternoon after your class? I could wait for you at school that day and we could ride home together. What do you think?"

"I have one of those." John reached in his pocket. "You need to see my teacher too."

"Stop. Let me make a cup of tea and change my clothes; then I'll look at my schedule. I promise I'll have these filled out and ready for you to take back to school tomorrow. Promise." Sometimes Britt was annoyingly conscientious.

"You, Britt, have piano to practice before dinner." While I put on the teakettle and took a half-thawed chicken from the refrigerator for dinner, Britt reluctantly went to the living

> **" *I was sure Mrs. Owen would understand the learning problems better than I did.* "**

room to practice. I heard the uneven scales from the other room. Within five minutes, Britt was back.

"Is my time up?" Britt asked.

"No, Britty, I'll call you when time's up. Do you have math homework tonight?"

"No, I don't have math on Wednesdays," Britt replied as she headed to the piano.

I opened my planner and sat down with the forms from the teachers. Yes, Tuesday of next week after my class would be fine, I thought. With any luck, I can see both teachers while I'm there. As I filled in the forms, I wondered to what extent Britt's problems with math and spelling were linked to the diagnosis of Turner's related learning disability.

Carol and Britt had been working on the assessment for the past month, but we still didn't have results, except that Carol was certain Britt had a learning disability. Britt could do her math orally—it was just a matter of getting it down on paper accurately. That would surely come with maturity, or would it? This conference with Mrs. Owen was just what I'd been waiting for. I wanted to talk to her about the problems and enlist her help. After all, Britt was wonderfully curious, lively, energetic, and articulate. I was sure Mrs. Owen would understand the learning problems better than I did. Surely she had had other children like Britt in her classes. Perhaps she would be an ally in solving Britt's problems in school. At the conference, I also wanted to talk to her about having Carol visit the classroom—the last piece of the assessment.

John came into the kitchen lugging a kettle filled with puzzle pieces. "I've been wondering

The Source for Visual-Spatial Disorders

Julie: Fourth Grade, continued

what happened to that," I said. "Where did you find it and what in the world is in it?" John climbed on a stool and poured the puzzle pieces out on the kitchen counter. "Britt turned the puzzles into a witch's stew again, so I need to put them back together," John said with resignation. He liked Britt's games of make-believe, but he also enjoyed puzzles. For Britt, puzzles always became anything other than a puzzle; I had never seen her try to put one together.

Britt was back again. "Now what time is it?" she asked.

"You've only been in there nine minutes total. Twenty-one more to go," I replied.

Downcast, Britt headed back to the piano. But she was back again in six minutes. "Okay," I said, "you can do another fifteen after dinner."

"Yippee," Britt said, "I'll set the table if we can eat in the dining room and have stories after dinner." I nodded and handed her the plates and silverware and asked her to choose some matching place mats and napkins. "Okay," she said, and then asked, "Where do the forks go?"

"Like I told you last night, the forks go on the left, and the knives on the right," I said.

"Oh, I always forget," Britt replied.

When we sat down to eat, some knives were on the right and some were on the left. "Britt, which is right and which is left?" I asked.

She held up her right hand and said, "left." Watching my reaction and her brother's response, she quickly reversed her hands. She didn't notice the positions of the forks or the knives and I decided not to mention it.

When we finished dinner, Britt asked, "Is it time to read now?"

"Sure," I said, "How about a Halloween story? The *Headless Horseman of Sleepy Hollow*?"

"Yea," Britt and John both said in unison.

I started to read. Britt dimmed the overhead light and listened with eyes sparkling in the candlelight. When I finally stopped reading, it was after 8:30. "Oh, I didn't realize it was so late," I said.

"How about my practicing?" Britt asked, looking for a way to avoid bedtime.

"You'll make it up tomorrow. It is time for bed," I said.

The next Tuesday, I dressed for the teacher conferences in my best black suit, black suede pumps, a white blouse, and red and black striped scarf. As always, I wanted to show Britt and John's teachers that I took my children's education seriously. And this time, there was also the issue of the learning disability.

Britt sometimes came home in tears because she had been unable to copy information from the board correctly. She was struggling in math and the situation was compounded because she often copied the problems incorrectly in the first place. Blaming herself, Britt had only recently shared her worries. Now it was time for me to talk to Mrs. Owen. I wanted her to understand that Britt was a smart girl with some unusual problems.

Mrs. Owen was sitting at her desk when I entered her classroom. "Hello," I said, holding out my hand. "I'm Julie, Britt's mother." I had met Mrs. Owen

> **" *I wanted her to understand that Britt was a smart girl with some unusual problems.* "**

at back-to-school night, but I wasn't sure she remembered.

"Yes, I remember you," said Mrs. Owen, shaking my hand reluctantly.

"You know," she said, "Britt is quite a good reader."

Julie: Fourth Grade, *continued*

"I know," I said full of pride.

"She's quite a bit ahead of grade level," said Mrs. Owen.

"Yes," I said, "Britt can read almost anything, and she has an excellent vocabulary. We read a lot at home."

"Yes, that's why it's so difficult to understand why she's so inattentive and makes so many mistakes in her work," Mrs. Owen hissed.

"I know math is difficult for her. According to the preliminary diagnosis of her learning disability, this is to be expected. We're looking for ways to help her. We should have a final report on her learning assessment soon," I was trying to stay composed, but I could feel Mrs. Owen's hostility.

Mrs. Owen interrupted, "She doesn't know her math facts. She should have known those last year. She doesn't even know 3 x 5 = 15, and I'm not sure she knows 3 + 5 either."

"She knows them quite well when we do them at home orally. She just has trouble writing them down," I said. "We're looking for ways to help her."

"Britt doesn't need anyone's help. She needs to develop better study habits. She needs to stay in her seat and do her own work. She's always asking for help. She can't seem to do anything on her own. She relies too much on being cute and clever. She's clearly a child who needs more discipline, clearer boundaries, not more help. She's gotten used to being dependent on others." Mrs. Owen finally paused.

"I'll talk to her about being more attentive," I said, feeling as if I was the child who had just received the scolding. "But some of her problems with spelling and math seem to be tied to her disability. Britt enjoys school even if she has problems in some areas. She likes learning all kinds of things. I'll speak with her about paying more attention to the details and staying in her desk, but I still think these problems may be tied to her learning disability." I could tell that Mrs. Owen didn't believe Britt had a learning disability.

Mrs. Owen looked away and began to stand up, signaling the end of the conference. "Just one more thing," I said, "I've talked to a learning disability specialist about coming to your class to observe Britt. The visit will be an important part of the assessment. We are really trying to find ways to help Britt succeed." Mrs. Owen looked out the window, not seeming to hear what I just said.

I continued, "Her name is Carol Stockdale, and she'll be calling you to set up a time to come to class."

Mrs. Owen nodded curtly, "Thank you for coming in, Mrs. Neff."

"Thank you for the conference," I replied. "I'll be in touch with you."

I left feeling uneasy. I also felt that I was to blame for Britt's problems. Maybe I had been too permissive. I was full of self-doubt. I would just have to try harder. Maybe my attitudes about what was important in school had somehow been conveyed to Britt. I lived in a world that revolved around reading and writing. Surely I would have been concerned sooner if Britt had not been able to read or write.

☙ ☙ ☙

> **"** *She relies too much on being cute and clever.* **"**

Carol: Visiting Britt's School

Britt's exceptional reading skill coupled with huge performance gaps confused me. Moreover, Julie had been upset after her school conference although she seemed to primarily criticize herself. I needed to see Britt in her classroom. Was she understood? Was she appreciated?

Early November is an effective time to visit school. Children are into the routine and the programs are well established. "Now do I have everything?" I asked myself as I left the house. Leaving the unanswered question hang in the air, I did not allow my mind to leave the tasks at hand: load the car, get to the classroom observation, go to the office, and plan the rest of the assessment.

This visit would be part observation, part conference, I reflected as I eased the car out of the garage. By the time my car had reached the top of the hill, I was fully focused on the small girl who waited in her Gig Harbor classroom. Britt's doctor was concerned about her anxiety level—tummy aches and headaches. What sort of burden was this ten-year-old carrying? Uneasy feelings had stayed with me for the last couple of days, but it wasn't the idea of visiting a new school. As the director of a learning clinic, I went to conferences every week. Something was different about this one.

Maybe it was the teacher's voice that put me on guard. I had called right after Dr. Medlar implored me to "get into that classroom."

When I had identified myself on the phone, Mrs. Owen had responded warmly. "Oh, of course, I was very welcome. Very welcome! She and her class enjoyed visitors."

My next words were, "At the request of Britt's doctor . . ."

Silence. I tapped the phone. The connection seemed intact. "Mrs. Owen?" Silence.

Then a curt, "When are you coming?"

Was this the same person speaking? "Mrs. Owen?" I repeated, stuck for other words

> **" *What sort of burden was this ten-year-old carrying?* "**

by the sharp change in her tone. "Yes, I asked when you are coming," she replied in the same surly voice. I let her select the day, requesting only that it include math period.

Today was the day for the conference, and 9:45 was the time we had agreed upon. Well, Mrs. Owen, in an hour I will be in your room. I wonder which voice will greet me. And I wonder if you know how anxious Britt is. Maybe, I thought hopefully, you are as concerned about her as her parents and doctor are.

I circled through residential streets toward the bridge. Sunlight shimmering on the snowy tops of the Olympic Mountains framed the mile-long span across the deep blue waters of Puget Sound. I breathed in the peace.

A few miles later, I eased the car into a "visitors" parking slot next to Carter Elementary School. Armed with my briefcase, clipboard, and questions, I followed the curve to the front of a long, low, brick and cedar building. All was silent as I entered the double doors and obeyed the sign "All visitors must report to the office." Soon a long-legged sixth-grade "office assistant" was escorting me through the hallways to Mrs. Owen's room.

My light knock elicited a voice charging Kevin to "greet our visitor." The door opened and Kevin, hair sharply parted and plastered down, whispered me into the room. Twenty-three curious faces turned toward me. Most were neatly groomed with a well-cared-for look. Mrs. Owen, in a navy shirtwaist dress, chalk in hand, announced to the class, "Mrs.

The Source for Visual-Spatial Disorders

Carol: Visiting Britt's School, continued

Stockdale is visiting" and to me, "You may sit at any open desk" as she turned back to the chalkboard.

On my way to an "open desk" in the next-to-back row, I took in the room: a wall of windows with a view of rhododendrons, trees, and the parking lot, green chalkboards across the front of the room, one of which was covered with writing, a large

> **"** *She looked up, blinking rapidly, a fixed little smile on her face.* **"**

teacher's desk in the front flanked by a student-size chair on one side and a large wastebasket on the other, and a table with six chairs in the space to the right of the rows of desks. The table was covered with a terrarium, plants, various-sized containers of water, folders, and two large magnifying glasses. I wondered if Elmo, the frog, would make an appearance today. Five rows of desks, six to a row marched precisely across the center space. Kevin slipped into the seat on my left while the seat to my right, labeled Jennifer, remained empty. Jennifer must be absent today.

Mrs. Owen was writing assignments on the right-hand board. Standing stiffly, she wrote without turning around until the listing of pages and numbers stretched in two columns from the top to the bottom of the board. She did not require eye contact with her students to maintain order.

No one in the class spoke. Each child was copying in a three-ring notebook, looking up every few moments to double-check the numbers on the board. Except Britt. She had looked back at me from her front row seat as I came in and lifted her fingers in a shy wave. Now she was looking at the board, pencil in hand. She put down a number, looked up, put down another and paused. She erased and rewrote and paused. Frowning, she pulled the page from her notebook, tore it in half, and put it on the corner of her desk. She stared at the board and began writing with her eyes fixed on the chalked message. Finally she looked down, sighed deeply, drew a line across her paper, and returned her eyes to the board. This time her lips moved as she repeated the words and numbers over and over, trying to hold onto them while she put them on paper.

Mrs. Owen announced the morning break, asking the students to visit the bathroom and return for announcements. Britt's hand waved in the air. "I'm not finished copying the assignments," she said.

Mrs. Owen sighed, "You will have to do it later, Britt. Put your things away."

Britt ducked her head as she closed her notebook. From my angle in the back, I could see the glint of tears in her eyes. She put her head down, appearing to straighten the materials in her desk, and used the moment to swipe her hand across her eyes. She looked up, blinking rapidly, a fixed little smile on her face. I could see her swallow as she filed out of the room with the others.

Before the students returned from morning break, Mrs. Owen moved to her desk and opened a textbook marked *Mathematics: Fourth Level*. She did not look my way.

I went over to the table that seemed to be an activity center and attempted to use that as an overture. "This looks really interesting. Are you studying life in fresh water habitats?" I ventured.

Mrs. Owen looked up and sighed a now familiar sigh. She came over, put her hand on the smallest bowl of dark, murky liquid and shifted it to the center of the table. "Oh, we get much more in-depth than that. These represent hundreds of hours of study. We do various water samples each week, and we test the water . . ." She described a yearlong program that sounded like it was suitable for much older children.

The Source for Visual-Spatial Disorders

Carol: Visiting Britt's School, continued

Concealing my reservations and biting my lip, I complimented her on a challenging and exciting program. "By the way," I added with my warmest smile, "call me Carol. Oh, and has the doctor shared Britt's medical condition? The spatial disabilities concerns—that's why I'm here—to help with this." I paused and waited before filling the silence with another comment, "When the assessment is complete, I'll send you the report. The description of Britt's learning patterns including the disability can help us address your classroom concerns."

Mrs. Owen paced over to her desk. I followed. She retrieved a manila folder from her top drawer, laid it down, picked it up, and finally, turned abruptly to me. Words spilled out. "Well, I received this letter from the doctor saying Britt has 'so-called' learning problems. But that doctor hasn't been in my classroom. That is for sure!" she blurted. Steps in the hall told me the children were starting to return.

I took advantage of the opening, "And what do you see, Mrs. Owen?" I repeated.

"I've been teaching for 26 years. I've seen learning problems. That's nothing new. I've got students with learning problems right now. Take Kevin, the boy sitting beside you. Slow—real slow. I usually have some of the other students go over things with him again." And, gesturing at the empty desk, "Jennifer—she's absent today which doesn't help a bit—she's way behind in reading. But Britt—Britt learns just fine—when she wants to," shot Mrs. Owen. She added, "She memorized all *fifteen* verses of a poem I assigned. Of course, I only expected each student to learn one verse. She claimed she did it at the bus stop. With Megan. She certainly knew them. So why doesn't she learn her math

> **" I've been teaching for 26 years. I've seen learning problems. That's nothing new. "**

facts? Can't be bothered—that's all." She paused, caught her breath, and frowned down at the folder she was tapping against the edge of the desk. Chameleon-like, she drew a smile across her face as she turned toward the students filing back into the room.

"Britt," she said in a soft voice, "Please sit back here in Jennifer's seat so that Mrs. Stockdale can observe you during math." She swung her arm in a broad semicircle directing Britt to the back of the room. From my angle behind her, I could see color rising in Britt's neck and cheeks. She swallowed hard as she ducked her head and shoulders to retrieve her materials from her desk. The flush had reached Britt's hairline by the time she set her math book on Jennifer's desk, slid onto the seat, and opened her notebook.

Mrs. Owen was announcing, "Today's assignment is on the board. Do both pages. There are no new skills in these problems so," she looked at me, then Britt, "you should be able to work independently." Mrs. Owen settled firmly into her chair, pulled a stack of papers from the second drawer on the left, and after one last survey of the class, began making rapid marks on the top paper with her red pencil.

I turned to Britt, who seemed even smaller in Jennifer's desk, which was one of the largest in the room.

"Perhaps you can show me what's going on," I whispered. "I haven't seen this particular math book before."

Britt brightened and pulled the text toward herself. "Well," she said, "This is it. She *always* puts the assignments on the board." She opened her notebook, "And sometimes—in fact usually—she erases them, so we have to have them written down." She looked up, her face sober and intense, with her eyes pinched at the corners. I looked

The Source for Visual-Spatial Disorders 39 Copyright © 2002 LinguiSystems, Inc.

Carol: Visiting Britt's School, continued

at what she had written in her notebook. Britt had recorded the first page and none of the specific problems assigned.

> " *Copying, erasing, recopying, counting—Britt stumbled through the math assignment.* "

The page had been erased repeatedly. I remembered the torn-up sheet of paper.

"Better finish writing down the assignment in case she erases the board," I suggested. She looked at the board, down at her page, back to the board, and back to the page.

"Would you read it to me?" she asked. My eyes went to the board and the several lines recorded there. I scanned the lines until I got down to: *Math-Page 69 even problems 2-14, Page 70 #1-5*. I read and Britt wrote.

Britt said apologetically, "She writes so much up there. I get mixed up. And then, sometimes, I even do the wrong problems."

"Well, you have it down now," I said in an effort to be reassuring. "Let's see what these problems are." Britt looked at the page number and back at the book. She propped it against the edge of the desk and started to flip pages looking for 69. She stopped for a moment and pointed to the number, repeated it, and returned to fanning pages. She stopped at a page that I could see was eighty-something. She paused and turned a page and another and another. She looked at me and fanned a few more pages. She was blinking rapidly. I slipped my hand between the pages and turned back to 69. "Let's see what you're working on," I added softly.

Multiplication problems covered the page starting with single digit, 5 x 9 = ___. "Okay, here we are. Do you copy them first or do you solve them as you go along?" I asked.

"Copy," Britt said. She chose a pencil from the ones she had brought, put it back on the front of the desk, chose another, lined up her paper with the math book, titled the page, put the date on the right margin, and sat back. "Um, will you check my numbers?" she said, ducking her head so I saw only the top of her head.

"Glad to," I replied. Holding her place with a finger thrust against the page, she began recording the problems. She wrote "2" for "5" in the fourth problem, but otherwise made no other errors on the first row. The 2-digit problems in the second row were much harder to copy. Britt went very slowly, but still reversed two numbers and substituted "6" for "9" on another. Solving them was another matter. Mrs. Owen was right; Britt did not know her times tables. She seemed to know the 2's and 5's, but the others were shaky, very shaky. She had to start at the beginning of the series and say each step to answer "4 x 4." I watched her lips form the words, "One times four is four, two times four is eight, three times four is twelve, four times four is sixteen." Copying, erasing, recopying, counting—Britt stumbled through the math assignment. When Mrs. Owen announced the end of math period, Britt had not yet begun the second page. With a deep sigh, she closed the notebook, stacked the textbook on top, and slipped out of her seat. "We get to read now; I better go to my own desk."

The noise in the room increased with papers shuffling, books thrust into desks or bumped off onto the floor by an errant elbow. Britt smiled back at me as she opened her reading book to the marker. Mrs. Owen was asking for someone to summarize the third chapter that had been completed the previous day. Many of the children looked down in an effort to be invisible, but Britt's hand waved energetically. "Oh, they went into the swamp and took a wrong turn around the big tree. They are lost and very, very frightened. And I think

The Source for Visual-Spatial Disorders

Carol: Visiting Britt's School, continued

they will have to make a signal fire . . ." came forth in response to a slight nod from Mrs. Owen. She wanted to go on telling what she predicted would happen next, but Mrs. Owen stopped her recital to urge the class to read silently. Quiet settled over the room.

Britt was oblivious to a buzzing fly, my observations, and Mrs. Owen's knit brows. She stayed that way until brought back by the announcement that lunch was imminent, and they had only five minutes to record their sentences about what they had read. Britt wrote rapidly for the entire time. She smiled at me. A bell announced the end of the morning.

After the children filed out, I casually scanned Britt's other books and her written paragraph. The books were the usual choices for the grade. However, the paragraph was not usual:

> "meg and Jocie are in the swamp becus the sing on the tree was wronge. Joice is woried but meg is note. In fact she is confidint that they will find there way."

Mrs. Owen looked at me and raised her chin. We were alone in the room. "Britt can spell correctly on a test, but you see how careless she is otherwise. You are looking for learning style—her style is careless."

I started to say, "But I'm not talking about style . . ." and stopped. Instead, I explained, "Learning disability' is very different from 'learning style.'"

Mrs. Owen ignored my words and continued, "She expects everything to come easily like reading does. I think her folks expect that too. It's just this divorce thing. She wants to be

> **❝ I had foolishly thought that she would welcome help. ❞**

everybody's baby and center of attention. She needs to buckle down, learn the math facts, and not expect special treatment."

Mrs. Owen walked me out of the room as she made this pronouncement. I was silent—stunned at the shallowness of her view.

I had foolishly thought that she would welcome help. After all, Britt was obviously intelligent and clearly needed assistance. How sad that Mrs. Owen saw only a nuisance.

I moved toward the parking lot brooding about Britt's tears, the distressed face, the frantic scramble to find the assigned page, the struggle to copy from the board, the gaps in math understanding, the sharp pleasure in reading, the rich flow of ideas, and the erratic, error-filled spelling. Sitting in the car with the keys in my hand, I did not see the beauty of the evergreens, but rather Britt's look of anguish. I could not agree that she did it on purpose. Britt was suffering. I shuddered. She had to face that classroom every day!

☙ ☙ ☙

Britt: The Rest of the Picture

Life outside fourth grade was composed of moments that left me unsettled and lonely and others where I felt confident and accepted. Pesky spatial problems intruded on classroom activities and normal kid activities. Learning how to ride a bike, play board games, cards, and sports are all supposed to be fun and a normal part of growing up, but they are difficult for anyone with a spatial learning disability. Being accepted is paramount to any fourth grader. I desperately wanted to do the same things that I saw all of the other kids doing—playing soccer, practicing piano, and roaming the neighborhood on a bike.

Like many normal American kids, I joined a sports team. The girls' soccer team associated with my elementary school, called the Ripples, maintained a winning record, at least during the time I played. Our success on the field happened with the help of several strong players and a coach that strategically placed me on the sidelines. With no talent for soccer, I became a misfit somehow shoved onto a winning team, not quite sure how it happened.

Dad seemed eager to introduce me to team sports and insisted that they would develop a sense of coordination and athletic ability not to mention toughen me up, which, in his opinion, I needed. Soccer appeared the natural choice because it was the first sport to be offered in the fall. I asked Mom if joining the team was really necessary, secretly wishing Mom would take my side.

"Yes, I think soccer will be good for you. You might make some new friends and fresh air never hurt anyone. Plus it will make your dad happy."

I begrudgingly acquiesced and joined the team.

Mom took me to the store to get all the equipment I needed for the first day of practice. Shin guards, cleats, shorts, and a uniform shirt with "Ripples" in large, fire engine red letters across the front.

I arrived at practice ready to play, or at least looking athletic. The coach explained the layout of the field and the rules of the game. She pointed out the opposing sides of the field and demonstrated how the soccer ball should only be touched with feet and knees and never with the hands or arms.

I took away three things from this session: get the ball in the goal at the other end of the field, keep the ball from going into your goal, and never touch the ball with your hands or arms. The rest remained a mystery to me.

During subsequent practices, I watched the other players for cues and dribbled the ball down the field just like the coach had asked. We also practiced passing the ball to each other.

The coach watched us all very carefully and assigned everyone positions on the playing field, sorting the players onto offense and defense according to their skills. The better players were picked to play offense, and everyone else stayed near the goal on defense. Falling under the latter category, I was to play halfback.

After assigning places on the field, the coach explained the different positions and the job of each player. As far as I understood, I had to stay near the goal. "Your job is to keep the ball from getting past you and keep it away from our goal. If you should get the ball, pass it to one of the forwards on our team, and let them take it back down to the other side of the field." Simple enough, I would stay by the goal and wait for the ball to come to me. That certainly seemed much easier than having to make the goals.

After gaining a comfortable lead of about ten points in the first half of our first game, the coach finally put me into the game. Once out on the field, I heard the coach yelling from the sidelines, "Get back behind the centerline and stay on our side of the field." Puzzled, I shrugged my shoulders, moved back, and wondered how my location on the field had changed. I quickly gave up trying to understand the strategy and followed the movement of the players and the ball across the playing field. After awhile, my mind began to wander. I began to think of the book I was reading, things that had happened at school, or stories I'd shared with the other neighborhood kids— anything to occupy my mind while I stood and dutifully waited for the ball to show up at my feet so I could boot it in the other direction.

The coach, tired of telling me to "get back on our side of the field" and "look alive out there," wisely placed me on the sidelines again. I never worried too much about being on the bench since I could just as easily daydream on the bench as on the field. Feeling guilty about keeping me out of the game, the coach tried to boost my self-esteem. In between the cheers of "Way to go!" and "Defense, defense!", she would look over at me and reassure me with "You're the secret weapon," and "We need to wait until the other team is not expecting us to put you in the game. That way we'll catch them by surprise and win the game." For the rest of the season, I sat on the sidelines during most games, content to watch the game from afar.

Unhappy at having to be at the games in the first place, I dreaded the moment when the coach motioned for me to head out onto the playing field. The strong forwards on our team dominated the field, thankfully leaving little opportunity for the defense to play. Watching the ball in play at the other end of the field, I secretly hoped that it would stay there, leaving me out of the fray.

During one game late in the season, when our team was up by sixteen points in the last half, the coach put me into the game. Forgetting that the goals had just been switched, I went to the wrong end of the field. The coach yelled for me to move to the other goal. I hustled to the other side of the field toward our goal. As I headed to my place on the field, the ball went into play. Trying to orient myself to my position rather than paying attention to the action on the field, I didn't notice that play had begun. The forward on our team kicked the ball to me since I was the closest to her at the time. The ball went up into the air in a perfect arc and headed straight for me with a direct hit to my stomach. I landed on my back in the wet, sloppy, rain-soaked grass. Unable to catch my breath and covered with mud, I headed back out of the game, defeated and frustrated. My soccer career lasted an entire two seasons, just enough time to placate my dad.

After trying softball during fifth grade, I opted out of team sports, having better success

> *" My soccer career lasted an entire two seasons, just enough time to placate my dad. "*

with swimming and gymnastics—both of which I could do on my own without having to orient myself spatially to a moving ball or other players. Dad never realized why I did not like team sports.

While I wasn't good at team sports, I was good at roller-skating, and roller-skating parties at the Skate King constituted major social events for our elementary school. The bright red carpet in the entrance clashed with the bright blue floor of the roller rink. Music blasted from the DJ booth and a disco ball made flickering patterns across the rink floor as it spun from the ceiling. Whirling around the rink in time with the music was exhilarating even when I had to slow as the floor became more and more crowded. Balance and stamina are prerequisites for skating, and I had both. Everywhere I turned, I saw

Britt: The Rest of the Picture, continued

someone I knew. I felt that I was in control, that I wasn't different.

Happy to find free time aside from school and soccer, my brother and I frequently recruited neighborhood kids to join our adventures. One afternoon, John and I raced up the long, steep driveway pushing our Schwinns through the gravel and onto the pavement as flecks of sunshine glinted off the handlebars.

Reaching the top of the driveway, we banded with a group of neighborhood kids and set up a game of cops and robbers. John and I, along with two other kids, were assigned the role of cops, and everyone else played bad guys. I kept up even though my bike still had the training wheels attached and my feet barely touched the pedals. We used walkie-talkies and chased the robbers down the street, hoping to cut them off and nab them before they escaped and retreated to the fort we had built in the woods behind our house. Finally the game moved into the make-believe town constructed of old logs in a clearing between the trees. Afternoon sun slowly fading into twilight signaled the end of the day's adventure and time to head back home for dinner.

Making up games and stories was fun and easy. Instantly imagining myself into a story line and taking on a different point of view placed me on an equal level with the other kids, allowing me to play as a full participant and even to organize and control the play. Suddenly I became a pioneer, a detective with a keen ability to track down thieves, or even a famous actress.

Every evening around dinnertime, Mom would remind me that I needed to practice piano. Mom had found a piano teacher who would come to the house once a week to give me a lesson. Since I liked music and I liked to sing, Mom thought that I might enjoy learning the piano. Plus, Mom had inherited a piano from her aunt and thought that it would be good to put it to use.

I remember the piano teacher as being old and cranky. She arrived equipped with a beginner's book full of recognizable tunes. The teacher asked me if I knew how to read music. Baffled at this question, I answered, "Yes." After all, I had seen music before, and I could read quite well. After an all too brief overview of sight-reading, she launched into the keys on the piano. Starting with middle "C" she went through the entire scale, going over flats and sharps as well as the regular keys. Then we played the first song, and I realized that there was more to reading music than just the words. The notes and lines were meaningless, simply dots and circles aimlessly placed on the page. Lost, I asked her to play the song first, and then repeated the notes by ear.

I reluctantly practiced every evening, never improving at the scales, and going over and over the lines on the page trying to pluck out the tune so obvious to the teacher. I tried reading the music through different methods, but all of them failed. I used the guide in the front of the book to try to name the corresponding note on the line of music. Before I was done, the entire page of sheet music was covered with pencil marks. After I had gone through and made notes, I placed labels on the keys of the piano with tape so I could match the notes on the page with the correct keys. This seemed like a good idea, but it did not help with the flats or understanding how long to play each note. I could not understand that a whole note

> " *Happy to find free time aside from school and soccer, my brother and I frequently recruited neighborhood kids to join our adventures.* "

The Source for Visual-Spatial Disorders

should be sustained longer than an eighth note.

Torture accurately describes the half-hour every day that I had to sit at the piano bench going over and over the scales. Every five minutes, I would check on how much time was left on the timer. Sometimes Mom gave up and let the practicing slide.

Every week the piano teacher showed up at our house ready for the lesson. I would begin to play the assigned songs, and without fail I would get halfway through the bar of music and the teacher would tell me with a horrified look on her face that it was all wrong. "Didn't you practice this week?" she'd ask.

My answer was always the same. "Yes, of course I did."

The piano teacher would take over and play the piece from the beginning and demonstrate how the music was supposed to sound, and then tell me to try to play the same part. I did, and by the end of the lesson, we had covered maybe one song.

"All right, I'm assigning the same music for the next lesson. Perhaps with a little more practice, you'll be able to finish the song without my help," the teacher would say.

After several weeks, I'd had enough of piano and began to look for any excuse to avoid practicing. After receiving a portable tape-recorder as a Christmas present from my mom, I decided that I would try recording half of my practice and then rewind the tape and play it back. I figured that this would allow me to go upstairs and read or play with the cat, and Mom would think I was at the piano. I taped fifteen minutes and then rewound the tape

> **" *Every five minutes, I would check on how much time was left on the timer.* "**

and fled upstairs. Suspicious because I had not appeared in the kitchen to ask how much time I had left, Mom went into the living room only to find the tape recorder sitting in my place. I heard Mom yell from downstairs and knew that I was in trouble. That marked the end of music lessons. I felt that I had failed, again.

Carol: Why Are Hard Things So Easy?

Moonlight made bulky shadows of the living room furniture, suggesting vaguely monstrous shapes. Cradling a mug of coffee, I relished the silence before the house stirred with the sounds of sons, husband, and dog. I needed these moments to review all that I knew about Britt. She was coming in to the clinic after school today for further assessment. I had found some articles in the local university libraries, but they added very little relevant information. Next week I would go to the Health Sciences Library at the University of Washington in Seattle. Meanwhile, I was intrigued by what she could do and what she could not do. How could we best use this next hour of assessment? Of course I recognized that the thinking required by math, copying, pattern matching, and finding hidden figures was similar and that reading and talking had a lot in common as well.

Now I needed to know more. Vocabulary and language were certainly areas to examine, but listening, speaking, reading and writing vocabularies could all be different. Somehow I felt Britt could do well at all of these. But that was only my feeling. I needed data. And language. Could she manage grammar and other skills? And what about memory? Did she have general memory problems or were they only in selected areas? Or were they really memory problems at all? Was there anything else I could learn about math and spelling? Clearly, these were problems, but how did she go about them? Was it that she had never learned or had she mislearned?

> **" She knew words far better than most children her age whether she was listening, speaking, reading, or writing the words. "**

And why couldn't she copy from the board if her eyesight was good? And why couldn't she find a specific page number in the book? A cold wet nose against my leg jarred my reverie. Peek wanted to go outside. Sighing, coffee cup in hand, I stretched. The day had begun.

My clinic day was busy with ordering more new books and testing materials, beginning the assessments of two new students both in high school, meeting with a college counselor about one of his students, and working on the syllabus for a graduate course I was to teach in a few weeks.

A few minutes before her appointment, Britt arrived with her cheerful "Hi" penetrating to my office as she came in the door to the waiting room. Soon she was in the low chair next to my desk telling me about her discouraging day that included criticism in math, a red-marked paper to redo in Language Arts, and bruises from being pushed down by a "big kid" on the playground. She cheered up when I put the large test booklet in front of her and said, "How about getting started by looking at some words?" Britt seemed to like every form of vocabulary test. Fortunately, the written test did not count off for spelling. She knew words far better than most children her age whether she was listening, speaking, reading, or writing the words.

We "talked math" long enough to determine that Britt could neither subtract nor add in her head. Given paper and pencil, she drew lines and counted them to get her answers. This worked unless she got mixed up and skipped a line or counted one twice. As a result she added "4 + 5" and got "10" while "5 + 4" produced "9." She did not notice the discrepancy. Counting was her only strategy. I decided to wait until her mother could be present to probe math more fully.

Britt needed to stretch and move around, so we walked over to a table stacked with a

The Source for Visual-Spatial Disorders

Carol: Why Are Hard Things So Easy?, continued

variety of toys. She picked up a plastic horse that was part of a farm set and rotated it casually in her hands.

"How about building a barn for the horse out of Lego blocks?" I suggested.

"Oh why not?" said Britt, scooping a pile of the plastic blocks toward her. She built awkwardly, making several changes as she went. At my suggestion, she left space for a door. She set down the finished barn that was four uneven walls, saying, "There, horse, just for you." To her surprise the horse towered way above the doorway. "Hmm," she said, "Well I just better get a smaller horse."

"Scale. Scale and space. Space relationships. Spatial relationships! She did not perceive the spatial relationship," said the voice in my head.

As we moved back to the desk, I reflected on Britt's struggle to find or create patterns. The Raven Colored Matrices that invited her to match patterns and the Visual Motor Test that asked her to copy designs had both been difficult for her. These were both in two-dimensions. Yet blocks were a three-dimensional task and that was hard as well. But she didn't seem accident-prone like people who didn't know their relationship to walls, tables, and things. How did she manage?

Tests during the rest of the session indicated Britt had very strong general language skills. She could combine sentences, compare similar and dissimilar sentence meanings, recognize categories of meaning, and identify correct and incorrect grammatical usage.

Near the end of the hour, I asked her to read a passage in a fourth-grade science book and

> " *'She did not perceive the spatial relationship,' said the voice in my head.* "

summarize the meaning. She read quickly, but stumbled when relating the content.

Interesting. The science content proved much more difficult to understand than stories or other print. "How could that be?" I wondered. The sentences in the science book were actually simpler constructions and tended to be shorter than in the stories. They often are straight facts: 'Plants breathe out oxygen.' 'Evergreen trees remain green in winter.'

Seconds ticked away as I mulled the behavioral contrasts. "One difference is no story line," I reflected. I knew story moved by using words to let the reader change images. Story and imagery were closely linked. Could the missing link relate to imagery?

Considering Britt's difficulty copying from the chalkboard and the problems she had finding fish hidden in a drawing, I wondered if she could track a line within a maze. I offered her several types of mazes and she chose a grid-like pattern in black and white. All of the choices were for six- to seven-year-old children. Britt was ten. We talked about the idea of drawing a line to guide the girl in the maze down a path without getting stuck in any dead ends so she can reach her puppy in the middle. Blue felt-tip pen in hand, Britt stared at the maze. She made several attempts. Then she laid down the pen and tried to slide her finger on a pathway. Nothing worked. The walkway, which seemed so obvious to me, was not evident to her.

Mazes were impossible.

As I closed the session with Britt, question after question pressed for answers to explain this—an "unusual learning pattern" didn't seem strong enough. I resolved to talk with Julie after Britt's appointment.

ಙ ಙ ಙ

Julie: Daily Life

That Tuesday in November, it had been raining all day. I looked at my watch discreetly so the student across the desk wouldn't notice. It was 2:55, and time to pick up Britt at school.

"Well, Amy," I said, "You know what to do—limit your thesis and use more quotes from the text. I think you'll be much happier with the paper."

"Thank you, Professor Neff," Amy said as she shoved the notebook into her backpack. "I'm so relieved to know that the paper isn't just awful."

"See you tomorrow," I said as I stood up, trying not to look rushed. As soon as Amy rounded the corner, I put on my raincoat, grabbed my car keys and briefcase, and left, closing the office door behind me.

I loved living in rural Gig Harbor, but on days like this with rain and rush hour traffic, I resented the 10-mile drive to Britt's school. But all of these afternoons that Britt had spent with Carol working on the assessment would be worthwhile if we could solve Britt's learning problems.

Britt was waiting for me in front of the elementary school when I pulled up. She ran to the car and jumped in. "Hi, Mom," Britt said.

"How was school today?" I asked.

"Okay. Kind of boring," Britt replied. "Guess what?"

"I can't."

"I need more lined paper."

"I thought I just bought you a package."

"You did, three weeks ago," Britt replied. "I'm almost out again, and Mrs. Owen will kill me if I run out."

"Okay, we'll stop at the store after you see Carol," I said.

"Could you do it?" Britt asked. "Remember, today is Wednesday. John and I are going out to dinner with Dad."

"Okay," I said. "I'll stop and get the paper."

I knew Britt was struggling with school, and I became more keenly aware of the problem after the report about the classroom from Carol. Still, Britt was often so cheerful that it was easy to forget how difficult things were for her.

"I hope Dad doesn't keep us out too late because I have to rewrite a story for tomorrow," Britt said.

"Wow, that's fast turnaround for a story," I said, thinking that I usually give my students as much as a week to rewrite a paper.

"I have to do it by tomorrow or she won't accept it. I have to do it tonight." Britt's eyes began to fill with tears.

"Do you want me to help?" I asked. "I'm good at revision."

Britt bit her lip, nodded, and gave my shoulder an appreciative pat as we turned into the clinic parking lot. The look on her face made me want to cry too, but I was afraid that would only make matters worse, so I changed the subject.

☯ ☯ ☯

> **"** *Britt was often so cheerful that it was easy to forget how difficult things were for her.* **"**

Carol: It's Not Sticking

Britt, her mother, and I were to meet for a few minutes at the end of the testing session. Julie wanted to ask a few questions and share Mrs. Owen's most recent message. I had my own questions for Julie. Tucking my notes into a folder, Britt and I moved to meet her mother in the waiting room.

After giving her daughter a hug, Julie handed me an envelope and we seated ourselves in the waiting room chairs.

"Mrs. Owen is concerned about a poem and some other issues," Julie began.

"Oh yeah, Mom, she made me stop," Britt impulsively broke in, "And the kids liked it, too. But Mrs. Owen was mad."

Turning to me, Julie added, "I talked with Britt about the poem she recited. Apparently she did get a bit too dramatic for Mrs. Owen but the fact is, she did learn all those verses. Thirteen or fourteen. And she learned them at the bus stop, singing with her friend Laura. I didn't coach her. She knows them now."

Britt interjected, "Want me to say them? I can."

"Not now Britt," said her mother, "I believe you know them perfectly." She looked at Britt and held out the car keys, "You have had a long day. How about getting the sack of apple slices and raisins I left on the back seat?"

As Britt left, I opened the note. Mrs. Owen's message began, "Dear Ms. Neff: Yesterday Britt put on a dramatic show reciting a lengthy ballad for the entire class. However, her mathematics skills have not improved." The note closed with, "Ms. Neff, to make you aware of how serious this is, you need to know that Britt is at the bottom of the class."

Julie looked at me as I read. "We simply must figure this out. The question I have is 'Why?' Why could she learn all those verses in such a short

> " *Why is learning math facts different from learning poems and other things?* "

time and without help? She works on the times tables night after night with everything we can think of. We have to admit, she really does not know them. Even the threes! Why? Why is learning math facts different from learning poems and other things?"

Just then Britt slowly pushed the door open and came in, handing the keys back to her mother, a plastic bag of fruit in her hand. She climbed up on the chair, settled against the back, and reached in for a section of apple.

"Britt," I began, "we are trying to understand why some things are so easy for you to remember and other things are so hard."

Britt slumped in her chair. "Like math," she said.

I thought for a few moments before speaking. "Britt, we all agree that you really like to read and that you know a lot of words—you have a big vocabulary, right?" Britt nodded. "Tell me about the last story in your third-grade reader—the one you ended with in June."

Britt brightened, "Oh that was about this family in an apartment house in Pittsburgh and the father lost his job and . . ." She described each person in the story. "The little boy had sandy hair—that means it was light brown . . . . The older girl got a paper route—she was the hero—she had braids and her name was Carla." Britt went on to describe the family members and their feelings in careful detail.

Smiling, I turned to Julie, "What do you think? Can Britt remember the story?"

"Absolutely," Julie replied.

Turning to Britt, I said, "I have a challenge for you. Can you

The Source for Visual-Spatial Disorders

Carol: It's Not Sticking, continued

pretend that the story happened in our town? That the same events happened—only here?"

Britt jumped to her feet. "Oh yes! They could live in the house that's for sale near my grandpa. And Carla could go to my school. But her dad wouldn't work in the steel mills; we don't have any . . ."

I interrupted, "That was great. You moved the story in a flash. Now let's try something else. You see six ducks sitting on a pond. Count them. Now three more come flying in and join them. They swim and swim. Two ducks waddle off. How many ducks do you see sitting on the pond?"

Britt was looking at the table where her right hand, pointer finger extended, was moving back and forth as if to follow the movements of the ducks. "Um," she said, eyebrows drawn together, "Five. Yes, I think five."

We worked through several other exercises as Julie watched. Britt could not count backward from 39. She could not count by threes or fours, and she got mixed up counting by twos.

Finally, I gave Britt a stack of cards with numbers written on them and asked her to arrange them in order of size from the lowest number to the largest. The small numbers under ten were easy. After that, Britt read each card and pondered at length before deciding their placement. Some cards were moved to new spots higher or lower than the first place she had put them. Britt sighed deeply. Finally, I took the remaining cards, saying, "That's enough of that." In ascending order the cards read: 6, 9 16, 33, 45, 43, 75, 57, 96, 92, 110, 117, 101."

"I think," I said, looking at Julie's puzzled face and Britt's

> **" You can picture many, many things very well—but not number. "**

discouraged one, "that you just helped us understand why math is so hard." I went on, "Memory depends on images. Britt, you did a wonderful job of making images to the story. You pictured the people and you pictured the events. You must be able to see the images before you can tell about them, and you can. You had no problem using your words not only to describe the people, but you could also change them. You could move them right here while holding onto the pictures. That skill is why you could remember the verses so well. I bet you could picture the events of every verse." Britt nodded. Then I continued, "It also helped that you sang them and the sentences sort of fit a rhythm."

"Now the math is different. You did not make stable images of the numbers. You really didn't seem to image the ducks—"

Britt burst in, "Oh I could see ducks, but not how many."

I smiled, "That's it in a nutshell. You can picture many, many things very well—but not number. There is even a big name for it. It's called 'internalizing the meaning of number' or 'linking quantity to numbers.' We aren't born being able to picture five things when we see the numeral five. In this example, it just means you do not picture six, so of course you cannot picture three more. You have the counting pattern memorized, but it is rote memory. To decide whether 102 comes before or after 120 or even 112, you have to count to discover which one you get to first."

Britt was examining the tabletop again. "And sometimes I have to start over lots of times," she said.

Gathering the cards, I tapped them on the table to even the pile and slid a rubber band around them. I smiled. "Britt has a wonderful memory for the images in a story or for events or, really, for most

The Source for Visual-Spatial Disorders

things. However, I suspect that she must see or experience things to image them—at least usually. And she has the language to translate those images back into retelling the story or describing something that happened. But she does not have accurate and reliable images for multiplication tables or, for that matter, for the number system.

One definition of verbal memory is 'using language to tell about images.' That really works for her in stories, but not in math. Because in math, she does not have the images to tell about. Now, before we are finished, we have to figure out what other types of images are hard to capture."

Julie interrupted. "So when Britt plays 'Little House' she has taken images from the story and translated them into her play. But when she sees a 'three' or a 'five,' she has no way to . . . well, hang on to the number?"

"Yes," I replied, turning to Britt.

"Well, Britt, this has been a wonderful afternoon," I said, leaning back in my chair and smiling at the small serious face. "Now we know where to begin. Images. Images for numbers. This should be an interesting trip," I added with a confidence I did not feel.

Britt looked dubious. Her eyes examined the feet of the desk while her finger drew circles on

> **"** One definition of verbal memory is 'using language to tell about images.' **"**

the gray Formica desktop. She looked close to tears.

"Well, let's get on with it," Julie interjected, pulling out her leather clad calendar book and standing up. "When is our next appointment?"

ಙ ಙ ಙ

Carol: Finding Answers

Mid-morning traffic was light on the way to the Health Sciences Library in Seattle. Often I enjoyed the drive, trying to absorb energy from the rush of vehicles. Today my focus was on Britt. After our second session, her learning pattern was becoming clearer. The distance between her strengths and weaknesses now appeared to be enormous. How could a child be so articulate and imaginative, so bent on ideas, and yet be unable to subtract two from seven or one from six? I mulled over her performance looking for points to research at the library. I had discarded drawing, mapping, mathematics, and visual perception as study topics after searching through these subheadings in two local university libraries. Six hours of reading fine print in reference books had gleaned several inches of journal articles. Yet they shed little light on Britt.

Many articles were on dyslexia, but Britt was not dyslexic. Dyslexic students had difficulty with the phonology of letters and words, and she certainly did not. Moreover, she could break words into syllables and language structure wasn't a problem. Bad spelling did not mean she was dyslexic. There are many reasons for poor spelling. No, Britt was not dyslexic.

Today I was looking at the medical perspective. Her health was good. But maybe medical research would probe physical causes of learning differences. I hoped so. This huge university with its medical schools and enormous research facilities was the place to start.

I had followed the directions from the attendant in the garage, but now I was at another intersection with hallways stretching in three directions. "The library?" My question slowed two young men, white coats flapping, as they passed.

"Follow us," said the taller one. I hurried to keep pace as we moved down the hall and through a set of double doors.

Stacks of books and journals on gray metal shelves nearly reached the ceiling. Gray tables and chairs were scattered at intervals amid the stacks. Many of those bent over books at the tables wore white coats, some with a stethoscope hanging from a pocket. A woman behind the reception desk absently tucked a strand of hair back into the coil along the base of her neck as she turned to me with a smile. "May I help you?" she asked.

A library search was the decision we reached after discussing the situation. The Rolodex in front of her spun as she sorted out the most appropriate journals and jotted them on the request form. We settled on key phrases, agreeing that "Turner's Syndrome" was the most likely, but that "chromosome deficit" was a possibility. I paid the fee and moved to a place at one of the tables. She would search journals in reserve and in back storage while I traced references in more current publications. Choosing the table with reader's guides, I thumbed through a huge volume to the label for Britt's medical condition. Sighing, I began compiling a list of articles, many of which were in journals that I had never heard of. Clearly, 1981 was a very busy year for medical publications.

Rubbing my eyes and replacing my glasses didn't help. My eyes objected to reading any more tiny print. I replaced the journals that were not remotely helpful and headed for the swinging doors to find some lunch. The student at the table by the door examined my bag

> *" How could a child be so articulate and imaginative, so bent on ideas, and yet be unable to subtract two from seven or one from six? "*

and gave me permission to leave. I chuckled at the sign "Gestapo" on his table. Many of these books and journals were irreplaceable and would disappear without controls. Myra, the "desk lady," thought she would have my other materials after lunch.

I headed for the sandwich machines. Soon I had a relatively fresh chicken sandwich and a cup of strong coffee. Sitting at a table, sipping bitter coffee from its Styrofoam cup, I thought about the articles stacked on a chair beside me and the others I had decided not to copy. Many of the authors were concerned with diagnosing Turner's Syndrome, with staining cells, with numbers and with consistency of incidence. Others focused on managing the physical matters that included heart valves and endocrine function. Only a couple of articles referred to learning. One group of authors concluded persons with this condition followed the population pattern in intelligence ranging from very intelligent to retarded. Well, of course. Britt was obviously very intelligent. So what! That did not explain her struggles.

Between bites of my sandwich, I mulled a phrase in another article, "a degree of space-form dysgnosia and a degree of dyscalculia may be characteristics" (Money, J. 1964). I considered the meaning. *Dys* means impaired or abnormal and *gnosia* is from the same Greek root as *know*. So *space-form dysgnosia* must mean an impaired ability to know space. That made sense. And *dyscalculia* means an impaired ability to do math. That is surely true. I shivered. Someone *had* seen this pattern before. The sandwich and even the coffee tasted better. Maybe the references Myra was tracking would tell me what to do about Britt's needs. I relaxed into the chair, elbow on

> " *I shivered. Someone had seen this pattern before.* "

the table, hand propping my head, and watched the stream of young people go by.

As I returned through the swinging doors, Myra smiled at me from her place at the desk. She pulled out a stack of journals from under the counter, each with a yellow card marking an article she had located. Eighteen more articles! I moved to a table, eager to see what Myra had found.

My experience in scanning articles helped me move quickly through the stack. The ninth article addressing Turner's Syndrome seemed to hold a clue, but not one I fully understood. ". . . *cognitive deficits are equally true of children and adults although the expression of these deficits may be influenced by particular stages of intellectual development*" (Garron 1977). This indicated that Britt's problems are unlikely to go away. ". . . *equally true of children and adults.*" So that's what Dr. Medlar meant. I reached for the next journal.

The sixteenth article, written by people from Children's Hospital Medical Center in Boston, looked at "Spatial and Temporal Processing" in girls with this syndrome (Silbert, Wolff, Lilienthal 1977). I read it, put it down to stare unseeing at the blur of print, and went back to read it again and again.

"*Patients . . . performed significantly poorer on tests of spatial ability than controls, but only on spatial tests requiring the integration of isolated elements as synthetic wholes or the remembering of spatial configurations which could not be verbally mediated. Patients also performed less well than controls on tasks of serial processing when the tasks could not be mediated verbally.*"

They were talking about performance on various tests involving space and time, and they knew Britt! Behind the complex language was the fact that their patients did poorly when they had to put together pieces to make something complete, unless they could talk it through. "Mediate verbally."

Carol: Finding Answers, *continued*

Yes, that meant talk their way through the task. Their students had problems remembering shapes and forms unless they could use words to portray the objects. And they had the same difficulty with "serial processing"—putting things in an order—unless they could use their own words to support the task. I remembered Britt's struggle with matching patterns or drawing designs, with copying from the board, with so many space and time tasks.

"Verbally mediated." Yes! Britt was at a loss when she could not talk her way through space and time tasks. That must be a key!

I shivered, remembering Julie's story about 10-month-old Britt. She had learned to label the plastic shapes in the sorting game at her grandparents' home. Britt would pick up a round piece, label it, "circle," and place it in the appropriate slot. I would wager that she talked her way through the task even then. Amazing! Not only was her language remarkable, but she also instinctively knew how to use it to succeed with the game. I remembered her method on one of the tests at the clinic. Looking at the square she was to copy, Britt said, "A square has four corners." Only then did she draw it. That made sense if she was using her language to tell herself what to do! I smiled at the idea that Britt was "verbally mediating" just like it said in the research.

In a few minutes, I had completed the copying, returned the

> **" When Britt could talk through the task, she was successful even with space and time. "**

last two journals, and was checking out of the library.

My elation lasted all the way to the interstate highway and through Seattle. My mind was on the idea that verbal mediation works. When Britt could talk through the task, she was successful even with space and time. I would help her develop skills to "verbally mediate." What a wonderful day!

As the miles went by, doubts began to creep in. The article suggested that some tasks could not be "mediated verbally." This included those that "*involved integrating isolated elements as synthetic wholes.*" Did that mean finding page 59 in a textbook? Britt didn't seem to understand the numbering system as a "synthetic whole," so page 59 was an "isolated element" that couldn't be nailed down. Maybe learning that was going to be harder. Maybe quite hard.

As the car moved south, the other category also started to worry me. What about "*. . . the remembering of spatial configurations which could not be verbally mediated*"? Was that spelling of sight words? And drawing? Lots of spatial configurations are difficult to translate into words. In fact most. What do we do about that? There must be a way. There must.

I didn't think about the complexity of visual imagery until I was nearly in Tacoma. How could Britt form accurate and specific visual images with this problem? So much of problem solving depends on images. Flexible, but specific and accurate as well. Concepts like *loyalty*, *integrity*, and *democracy* needed this kind of thinking. The sun had dropped behind the Olympic Mountains and dusk was settling. I turned on the windshield wipers as the first gusts of rain hit. The fine weather was leaving, but it would be back. We could count on that.

ಙ ಙ ಙ

The Source for Visual-Spatial Disorders

Julie: Having Answers

In early December, Carol called to say that the assessment was complete. I scheduled a meeting with her for the next week to go over her results and to decide what to do next.

Again full of hope and dread, I took my place across the desk from Carol. We were both focused on the manila file on Carol's desk marked BRITT NEFF. Perhaps I should have been more anxious about these results, but I continued to believe that if we could understand Britt's learning problems, we would be able to address them. Perhaps I was just buoyed by naïve optimism.

Carol opened the folder and explained that Britt had a spatial learning disability. Carol said that while Britt's capacity for language was strong, as evidenced by her ability to speak and read, her spatial system was severely underdeveloped.

I listened in disbelief. What could this mean? I had never heard of such a thing. Dyslexia, yes—many of my students over the years had had this problem, but a spatial disability? Carol explained that not very much was known about this disability. She had been combing the libraries, even the medical school library, to try to find out more about how the brain manages spatial tasks. She said that at this point she only had pieces of the puzzle, but she hoped that when she put all the pieces together, they would help explain Britt's spatial disability. Carol explained that Britt has no concept of number; that's why she has such a hard time with math. "That's why math facts just don't stick in her brain," Carol said.

Carol explained that Britt's problems with spelling were similar in that recalling the order of the letters is a spatial task. "Spelling," Carol said, "is a visual memory task, and Britt

> **"Spelling is a visual memory task, and Britt doesn't have an accurate visual memory."**

doesn't have an accurate visual memory. She has no images to rely on. So she spells things phonetically, and that means she often spells things incorrectly. Many everyday tasks are spatial."

Carol continued. "Telling time and drawing are spatial tasks too, and both are difficult for Britt. And puzzles are impossible for Britt, right?"

"Yes, they are impossible for her to put together. She plays with a puzzle by turning it into something else, like putting all of the pieces into a kettle and calling it a 'witch's brew.' "

I listened intently, trying to make sense of what Carol was telling me and struggling for the questions that would help me comprehend. "But why does she read so well? Why is she so articulate?" I asked.

"Well I don't know for sure," Carol said, "but the language system is different, and Britt uses her strong language to compensate for her spatial disability." Carol paused. "She's smart; she's very smart. That's why she has been able to get along as well as she has. As she reads, she has narrative to help her build visual images, but she has no images for things like numbers," Carol said. "I think that's why these problems didn't show up until third grade. When the math and spelling got harder, she found her coping devices weren't adequate anymore."

Carol went on. "In fourth grade, the teacher stressed copying from the board. Britt had difficulty with this because of inaccurate visual memory. She just wasn't able to transfer what she saw on the board to her paper."

Sitting quietly, I began to let the enormity of the problem sink into my heart and mind. I no longer felt that understanding the problem would allow me to solve it. "These

The Source for Visual-Spatial Disorders

Julie: Having Answers, continued

seem like insurmountable problems," I said, as I worked at maintaining my professional façade. "I can't imagine how she can live a normal life with all of the things she can't do." I was only beginning to comprehend the implications of what Carol had just told me.

"Well, we can work with her here at the clinic to teach her how to cope with a variety of spatial tasks. But I need to do some more thinking about this, about what would help her the most," Carol said thoughtfully. "Let me think about strategies, and I'll call you after the holidays. This isn't a good time of year to start kids on a new program anyway."

"Okay," I said, still groping to understand. "So the fact that she turns puzzles into 'witch's brew' is the way she makes sense of the puzzle? She builds a narrative around the puzzle pieces, so she is in charge. If she controls the play, then she can get everyone else to play on her terms?" I paused.

"One more thing," I said. "I've been meaning to ask you about this. A few weeks ago, I took Britt, John, and Britt's friend to see the University's production of *Dracula*. The kids like to be close to the action, so we sat in the front row. When Dracula arrived on stage, Britt jumped into my lap, and Laura and John soon followed. We were quite a sight. This student play

> **" Could Britt manage high school? How would she function as an adult? "**

was scary, but not that scary. But here's the funny thing—the week before, Britt's dad took the kids to a horror movie. It was much scarier and more realistic than this play, and she wasn't bothered by it at all. In fact, I was concerned that her dad had taken them to this movie, but Britt just said that it wasn't very scary. Amazingly, she had almost no response to it. John, on the other hand, had nightmares about it. Maybe it's just that he's younger. I just don't know."

Carol listened intently. "That is interesting," she said. "Very interesting. Let me think about that too."

"Thanks, Carol," I said as I stood up. "I'm really looking forward to hearing from you and getting started with something that might help Britt with the math and spelling. Merry Christmas. And thank you again."

I left Carol's office and headed back to the University to grade final papers. Yes, I had some answers, but I was stunned and overwhelmed at the severity of the problem. As I pulled into traffic, I couldn't help wondering what other tasks might be spatial in nature. Could Britt manage high school? How would she function as an adult?

Still I had to believe there was hope. I had to have faith that Carol would find a way to help Britt, and I had to have faith that Britt had talents beyond her disability.

ಇ ಇ ಇ

The Source for Visual-Spatial Disorders

Hope

*"Hope" is the thing
with feathers—
That perches in the soul—
And sings the tune
without the words—
And never stops—at all.*

Emily Dickinson
1830-1886

Carol: Inner Space Is Not Outer Space

Julie's questions at the conference confirmed just how difficult it was going to be to explain why Britt had problems with spatial concepts. This information was simply not something most people knew. I mulled over the challenge on this Saturday in December as I set about my weekend routine.

Saturday morning at home meant vacuuming and laundry, tasks that left plenty of time for thinking. Joel was helping his father with yard cleanup. With both older brothers in college, there was plenty to do.

As I was shoving his jeans into the washing machine, I thought about using Joel to test out language. After all, he and several of his friends had helped me before when I needed to do some trials on a new memory assessment. Being paid in pizza was part of the attraction, but they had also proven to be interested subjects.

I had already tried out explanations on colleagues with mixed results. Since testing Britt, visiting her school, and then gathering the research information at the "U," I had sought clear descriptions that used the words *spatial*, *spatial learning*, *spatial tasks*, or even *spatial information processing*. Clearly, this was Britt's struggle. But even her mother, who desperately wanted to understand, found the ideas unfamiliar—even confusing. How could I convey Britt's needs and find solutions if I could not explain the fundamental concept in simple English?

Not only were the ideas novel, but also the words meant something else to most people. Perhaps my youngest son could give me practical feedback. Joel shrugged his okay to my request. I told him that I was trying to help a girl with spatial learning problems.

"Space means orbits and astronauts to me, Mom," said Joel, as he scooped a large bite of chocolate ice cream into his mouth. "Nothing to do with school, that's for sure," he added.

Even though I knew that conceptualizing spatial relationships was fundamental, Joel's words underscored the difficulty of expanding his meaning for *spatial*. Frowning, I took another sip of coffee, balanced the mug under my chin with both hands, and tried to think of words to explain how *spatial* and *learning* went together.

I couldn't start with anatomy. Joel really didn't care that his brain had two hemispheres. While the researchers were very concerned about theories of problem solving, Joel wasn't. He needed examples, personal examples, which were real to a 14-year-old. As the youngest of three athletic sons, he had lived in a world of sports and vigorous activities.

"Can you tell me the names of all the players on your baseball team from last summer, Joel?" I ventured.

Eyebrows raised, he looked at me as if I had lost an important part of my mind. "Gee, Mom, Terry Andrews was catcher, Dave Teryll was backup, Mel, Richard, and Eddie pitched, Bill Tminszke and Adrian Wiley alternated on first and . . ."

"How are you figuring it out so you don't skip anyone," I interrupted.

"Mom, I'm just going around the field. But you didn't give me a chance to get to third where I play."

"Joel, you were thinking spatially. You just pictured the field and the players in their positions," I said.

> **"** How could I convey Britt's needs and find solutions if I could not explain the fundamental concept in simple English? **"**

The Source for Visual-Spatial Disorders

"Well, of course," he replied.

"As captain, you helped Coach Adams by filling out the league roster," I said, adding, "Was that in the same order?"

"No," Joel said with a questioning look on his face, "that had to be in alphabetical order. You know Aaronson is first on every list and then Andrews, but I don't know the rest."

"Okay, Joel, that's another kind of thinking—another kind of information arrangement—one that follows a specific rule of sequence—and it's different from the spatial arrangement," I added, "Can you feel the difference?"

He looked dubious, "I dunno," he said, eyes on the ice-cream carton I was cramming back into the freezer. "Alphabetical is alphabetical, that's all," he added with a shrug.

I needed another example from Joel's experience. "Okay, think about all the model airplanes you boys built. You and Brant would take one look at the picture, dump out the pieces, and know how to assemble the plane. Unless you lost a few struts along the way, the planes looked great. You could hold the design in your head and assemble the parts into the finished product. You built according to a spatial concept in your head."

"Yeah," added Joel, "and they flew too. Especially the ones with the little motors. But everybody can do that, Mom."

"Remember how Allen built his planes? He might look at the picture, but he carefully got out the assembly instructions and went through every step—1, 2, 3. His planes flew too, and he never had any parts left over. He used a more linear problem solving approach. You and Brant used a spatial problem solving approach. Two kinds of thinking. Do you see a difference?" I asked.

"Well," Joel replied, "I could do it Allen's way and plow through all the instructions. I just don't want to."

"Yes, I'm sure you could. But what if you had a glitch in your problem solving abilities and could *only* do it by reading all the steps. What difference would that make?" I asked.

"Um," he looked out the window. "I'd hate that and I probably would never touch another model—wouldn't be fun," he mulled.

We talked about the choices he made that centered in spatial thinking. He could picture where everyone was on his soccer team in relation to a play even when they were running full speed up and down the field. He knew how to get to the swimming pool even when the main road was blocked by construction and he had to make many, many twists and

> *Most things can be solved more than one way. But one choice is often more efficient if your system lets you do it either way.*

turns on unfamiliar streets. Of course he could have had instructions written out that said, "Go down Grandview to 19th, turn left, go to Seaview, turn right" and on and on.

Talking around the spoon in his mouth, Joel replied, "And that would be more analytical, right, Mom?"

"Right," I said. "Most things can be solved more than one way. But one choice is often more efficient if your system lets you do it either way."

"There's more to it than picturing routes, distance and direction or whatever is in your head, Joel. You also have to picture—maybe *feel* is a better word—the systems we use. Measurement, for instance. How long is an inch or a yard? How heavy is a pound? Understanding and picturing these kinds of systems is part of spatial thinking," I said, wondering how long Joel would be

The Source for Visual-Spatial Disorders

Carol: Inner Space Is Not Outer Space, continued

willing to think about this "teacher stuff."

Joel reflected as he slid down in his chair. "Mom, I just thought about something. Once when I was in fourth grade, Carl and I were playing at his house flying airplanes—just paper ones—out the upstairs window."

"Anyway, when we went down to get the airplanes, the door shut, and we got locked out. Boy, were we scared because we had turned on water in the tub to try out this bike tube we had patched. Carl said, 'Let's use the ladder and go in the upstairs window.' He came out of the garage with this dinky little 8-foot ladder that wouldn't come anywhere near reaching the window. Only he didn't know. He got really mad when I said it wouldn't work. He couldn't just look at it and tell. He does a lot of things like that."

"You're getting it! He couldn't estimate the distance or compare it to an 8-foot ladder. Did you two flood the Sandken's house?" I asked.

"Nah. Carl's brother came home with his key, so no one ever knew," added Joel.

"How does Carl do in math class?" I asked.

"D's. He hates math. Says it's for nerds," replied Joel.

"Sometime I'll tell you why math is the most demanding of spatial thinking of all school subjects. But lots of other

> **" Telling time, estimating time, grammar, and even punctuation is spatial. "**

things are spatial as well. Telling time, estimating time, grammar, and even punctuation is spatial. But you don't need to hear all that now," I offered.

"Right, Mom, I've got things to do. But you know, it's kinda interesting to think that your brain has different gears," he commented as he angled across to the dishwasher with his empty bowl.

"I think it is more like different roads," I amended, looking at Joel stretch upward in the doorway with his hands pressing the top of the frame. "You can often get there by both roads, but one is a lot more direct."

"You know, Mom," said Joel pensively, "I think Brant is really super at spatial stuff. Remember how he could make those planes with motors and run them by remote control to do all sorts of things. He could dive them down so exactly that he could clip weeds or even tall grass stems and then land them on the roof just as smooth. He was lots better than me or anyone. His buddy, Doug, practiced and practiced but still cracked up every time he did anything at all tricky."

"You are probably right, Joel," I said, thinking about Brant's umpteen construction projects. "But coordination and motor control are also important. We haven't talked about them yet."

"It's okay, Mom. That can wait," he added giving me a quick grin.

I watched him as he headed for the door. He doubled back to pick up the baseball bat propped in the corner, tossed it into the air, and caught the knob on the bottom against the flat palm of his hand. He headed for his downstairs bedroom, balancing the bat on his left hand. "How's this for spatial, Mom?" he called over his shoulder.

✧ ✧ ✧

Spatial - math, spelling, time, estimation, grammar, punctuation

Carol: A Window Is for Seeing

Britt's assessment was officially completed in December. During the early months of the next year, Julie and I had to face the fact that Mrs. Owen did not or would not recognize Britt's needs. The teacher remained aloof and critical. Julie provided extra help at home and made sure that Britt's days started and ended with hugs. Britt was under too much pressure to add clinic instruction to her week, however much she needed help shaped to her special needs. But by April the time had come to plan.

During one of our phone consultations, Julie and I agreed that knowing about the problem was commendable, but doing something about it was what mattered. I put aside time to review what I had learned from research during Britt's assessment process and to put forward ideas for helping her. While the medical research reports explained why and how Britt struggled, they did not tell us what to do to help her become successful in her problem areas. What could we do? Britt had vividly demonstrated her needs and her strengths during testing. She needed to understand space and I wanted to help her do so.

Beyond talk, change had to be visible to Britt and to those around her. Our experience with Mrs. Owen made this plain. We had to make Britt's needs visible so that we could measure progress for her and for us.

I realized as I planned a class for Britt, that doing something with space meant building or drawing. I recalled her struggle during testing. Britt had looked a little doubtful when I told her that I wanted to do some experiments, but that I was not trying to make her into an artist. "That's good," she replied, "because it would never work."

First, I let her select two items to draw from our clinic junk box. Her drawings were vague blobs. Perhaps it was intimidating to draw from a model? We had then shifted to drawing from memory and I asked her to draw something she recalled from home or school. She chose her desk. Again the rough square she drew could have been anything. We had tried numerous approaches all with the same result. Her drawings looked like those of a young preschool child. Being able to clearly draw objects—making them flat on paper—was difficult for her. In fact, she couldn't do it. That led me to wonder if it was

> **" We had to make Britt's needs visible so that we could measure progress for her and for us. "**

transferring a three-dimensional object to two dimensions that was so hard.

Perhaps, I reasoned, reversing the process would be less troublesome for her. Possibly going from the two dimensions of a picture to three dimensions would be easier. The modeling clay we used was new and soft.

I remembered Britt choosing the picture of a circus elephant to model although she looked very dubious about the task. She had rolled little balls of clay on the table until she had made seven or eight of them.

She chatted all the while. "I like elephants—at least to look at. They are kind of funny with that long trunk . . ."

Next, she tried to make a big ball for the body of the elephant. She stuck some of the balls onto the "body" and then looked at me saying, "I don't know how to do the head." She tried to stick another piece of clay onto the body in several ways. Finally she sighed and stopped. Evidently, going from a two-dimensional picture to a three-dimensional form was no easier for her than the other way around.

As I reviewed her efforts during assessment, it became

The Source for Visual-Spatial Disorders

Carol: A Window Is for Seeing, *continued*

apparent that Britt's problem was fundamental—she did not "know" how things looked. Visual perception was a major problem. And in many ways this lack of knowledge wasn't apparent to her. At this point, my planning took on direction. Maybe, just maybe, there was something we could do to show her.... No, that's not right. Not show her. She needed to discover how things looked and to construct images for them in her memory—useful flexible images. So many parts of learning are built on this knowing. Perhaps if we set up the right situation, she could discover.... My mind teased the notion, like a tongue probing a cavity in a tooth even though each thrust hurt. How could she discover? How could she know?

During these days and weeks that I planned Britt's class, many of my daily experiences triggered links to her and her needs. Like when I noticed the bicycle one of my sons had left leaning against a post in the yard. In my mind's eye, I "saw" Allen wheel the bicycle to the storage shed behind the house and resolved to remind him to do so. My mind went back to Britt. Something, some idea was pushing at me. My mind ricocheted to Allen, the bicycle, and the images of it leaning against the post facing away from me. In a flash the image reversed, and I could see the front and top as Allen wheeled it toward the storage shed. At the same time, I could see it inside the building. Britt couldn't do that. She could not

> **"** *Something was there—an idea began to gnaw at me just beyond my grasp.* **"**

manipulate images. At least not some of them. She didn't "know" objects the way most people did. That's why she couldn't draw them. I didn't understand how her images were different—but they were.

Having taught art and studied children's development in and through drawing, I knew the stages they went through. But I also knew they could learn about shape by drawing. Perhaps Britt could if we could figure out an entry for her.

Understanding Britt's spatial deficiencies had helped me see spatial problems in other children. Helping Britt might supply a key for helping others. She had many more strengths than most children, and her problems were narrow even though they were severe. But what could be done with a bright kid who couldn't picture shape or number? I reviewed what Britt could do so well and what she seemed unable to do at all. Something was there—an idea began to gnaw at me just beyond my grasp.

During late winter and early spring, other children with spatial problems came to the clinic. Would I have diagnosed their problems if Britt hadn't given me the insight? Probably not. Tom scored below percentile scales on tests that asked him to perceive form and shape or to match designs. Jeremy could do none of these tests, nor could he recall how to write letters and words. In fact, he was nearly unable to write. Like Britt, these students had baffled their teachers and parents. Space did not exist for them. I started a list of children with spatial problems. Nine names were on it from just a few months of clinic work. How many students went undiagnosed? How many had I looked at, but not noticed before this?

Ideas for helping Britt and other children with related problems tantalized me. They simply could not duplicate the specific forms or shapes—the spatial qualities of the objects—but maybe we could find a way for them to do just that. From some distant art history class, I recalled the experiments done during the early Renaissance in trying to analyze perspective by tracings on glass. I had experimented with the concept of such a "tracing window" when I taught art. In fact, several of the students had traced

The Source for Visual-Spatial Disorders

Carol: A Window Is for Seeing, continued

objects through the glass door in the classroom. It had seemed to help them draw more accurately.

Maybe some sort of window drawing would help Britt understand forms by tracing them. Couldn't hurt. And it might help. All we needed for the experiment was a piece of plastic in a frame that stood up so we could put objects behind it. I reached for a sheet of scratch paper and started drawing my idea.

In a week, the plastics company had translated my sketches into what I call a **Learning Window**. The two vertical panels of plastic measured 18 x 24 inches and were sandwiched together into a stand. Objects could be placed on a table and viewed by a person looking through the window. While looking at the object on the other side of the window, the person could then trace the object onto the window using an erasable pen. The panels were also removable so that a drawing could be copied on the office copier.

Perhaps this was Britt's tool for translating objects from three dimensions into a two-dimensional format. The idea seemed almost too simple to work. My plan was to use her rich language to describe the experience—but most of all, to "draw" in a way she had never done before. Now we could shape specific images with her.

We could make that crucial three-dimension to two-dimension transition that seemed so elusive for her.

ಬ ಬ ಬ

Julie: The End of Our Rope

On a Saturday morning in late May, haze covered Hale's Passage as I sat alone at the kitchen counter with my second cup of coffee. Finals were over at the University, Britt and John weren't up yet, and I finally had time to think—and to worry.

As fourth grade was drawing to a close, Britt's problems in school had been on my mind. I understood she had a learning disability, but I had no idea what to do about it. Nothing I tried seemed to work, and she was falling further and further behind in the subjects that were difficult for her.

Britt still did not know her times tables, and she still had trouble with long division. Her spelling was poor. We worked every night on the times tables—singing them, making rhymes and rhythms with them, writing and rewriting them. Even if she knew them perfectly in the evening, by the next day she couldn't write the answers on a piece of paper. Why couldn't she learn them?

We had the assessment, but what did it mean? What did it really mean in terms of how Britt learned and how I could help her learn? Nothing I had tried had worked; not anything in my years of teaching had prepared me for her problems. And Britt's teacher didn't have any interest in trying to understand or to help. Why couldn't the school or the teacher do something to help Britt? Was it that they didn't want to help or was it that they didn't know what to do any more than I did?

This year had been awful; bad for me and especially bad for Britt. Because of the emphasis on her disability, she was losing confidence in her abilities, and her other schoolwork seemed to suffer because of it. Much of the joy she had had in learning was beginning to disappear. As the year had passed, I saw my bright, lively daughter who should be, I was told, at the top of the class, sinking to the bottom.

I had been so frustrated I had felt like slapping her, but I didn't. I wanted to say "Sit up straight and learn this math!" but I didn't. As the days lengthened and the evenings grew warmer, I longed to go for a walk or sit on the deck and read a novel, instead of doing math drills. I wanted a break from this constant problem.

I had begun admitting to more of my friends that Britt had a learning disability. I suppose I had been looking for support or answers when I discovered that Britt had a long-term problem. Their responses were always the same: "Oh, that can't be. Britt has always been so bright, so articulate, such a leader." I assured them that it could be, telling them that she didn't know her times tables, she couldn't do long division, she couldn't spell with any predictability, and she couldn't follow directions. I told them how we had tried and that nothing had seemed to do any good. I told them about the diagnosis of a severe learning disability that was not dyslexia. I hoped that they could tell me what to do.

> **"Nothing I tried seemed to work, and she was falling further and further behind in the subjects that were difficult for her."**

Family members had a similar reaction, but they seemed to be in complete denial: "I'm sure she'll grow out of it." "She's so clever and cute she can get by without knowing her math facts." "Britt will always be Grandpa's precious little girl." "Mary Beth's grandson had dyslexia too, and he's just fine now. He's got a good job at the mill, a nice wife, and the cutest little baby. You should just see how cute the baby . . ." I was furious at their lack of understanding. They were unable or unwilling to help Britt or me. Maybe I did not want help, but I did want understanding. I wanted to scream: "This is a

Julie: The End of Our Rope, continued

problem! 'Cute' doesn't count!" And yes, a big vocabulary is nice, but it doesn't count now either. What are we going to do next year in fifth grade? And the year after that?

Britt seemed increasingly anxious and upset. She looked for excuses to stay home from school. During the spring, we wanted to do some work with Carol. But since Britt's teacher remained unwilling to listen, much less provide reasonable accommodations, we had to focus on simply getting through the days and weeks. Britt was now far behind the rest of her classmates in math and spelling.

Something had to be done about Britt's problems. If I didn't do something, who would? Carol seemed to be the only one who understood the gravity of the problem. On Monday morning, I would call Carol, tell her about my concerns, and make an appointment to talk about what we might do over the summer to help Britt. Clearly, the math drills hadn't helped, but I had to do something, and I couldn't think of anything else.

My musing was interrupted by the canned laughter of the Saturday morning children's shows. The kids must be awake, I thought, as I poured another cup of coffee. Looking at the boards that needed to be replaced on the deck, I wondered how much it would cost to get Britt the help she needed.

By 3 p.m. on May 25th, I was sitting in Carol's office. Carol, in her reassuring professional voice, told me that she had been thinking about my call and

> **" *I was filled with relief that this just might work to address the problems.* "**

Britt's problems. She also said that she knew of a few other children with similar problems and was thinking about a summer class that would help them with their spelling and math. She wanted to know if we would be interested in the class for Britt. The cost would be about $300 for a two-week session that would last from 9:00 to noon, Monday through Thursday.

Of course, I was interested. I was also desperate. But I told Carol I would have to think about it because of the cost. There were a million things that needed to be repaired around the house, and a $300 class for Britt had not been in my budget.

That evening, I called Britt's dad to see if he could split the cost with me. He agreed that money was a problem, but we really didn't have any choice. Britt needed this class. The next morning I called Carol right after Britt and John had left for school. Britt would enroll in Carol's summer class. I was filled with relief that this just might work to address the problems.

That afternoon, soon after I heard the school bus grinding up the road, Britt and John bounded through the back door and into the kitchen, throwing their book bags under the counter. John was retrieving the milk from the refrigerator while Britt moved the cookie jar into easy reach. I handed them each a glass for milk.

"Britt, I've got good news for you!" I exclaimed. I was happy about the summer class.

"What?" she asked, reaching for a cookie.

"I talked to Carol yesterday, and she said that she is going to teach a summer school class that will help you with your math and spelling. Isn't that great?" I said. Before I had finished my sentence, Britt's face had fallen. She put down the cookie as her eyes filled with tears. She seemed to shrink in the chair.

"Oh, Britty," I said. "I'm sorry. I know what a hard time you've had." I went around the

The Source for Visual-Spatial Disorders

Julie: The End of Our Rope, continued

counter to put my arms around her. "But working with Carol will help; I know it will."

"You *don't* know," she snapped. "You don't know at all."

"Well, maybe I don't know in the way you do, but I do know you want to be successful in school. You want to be as good in spelling as you are in reading. And Carol is so nice. I don't think this will be bad at all. It's only for two weeks."

"All day, every day for two weeks. It's going to ruin my summer," Britt replied, her tears turning to a firm pout.

"Can I go to summer school and take math?" John asked. He enjoyed reminding Britt that he was good in math.

"Shut up, John," Britt said.

"No, John," I said. "You're not going to summer school in anything but swimming, maybe. Now how about going outside to play?"

"But . . ." John started.

"We'll talk about this later. Take a cookie and work on the fort," I said. John, then a second grader, had been in a second/third grade split. He took pride in being near the top of the third grade class in math, and he seldom missed an opportunity to remind his sister of this, often offering to help with her math homework.

This time, though, John knew from the tone of my voice and from the look on Britt's face that it was time to find something else to do.

Britt jumped down from the stool and walked deliberately out of the room. A few moments later, I heard the door to her room slam shut.

ಬ ಬ ಬ

Britt: "It's Good for You"

As fourth grade ended, Mom informed me that I would be taking a class over the summer with Carol to improve my math and spelling skills. Images of going over endless math problems for an entire summer sprang into my mind.

Thinking that I would have to spend three months agonizing over thousands of numbers that never added correctly in the first place made my stomach churn. Frustrated, I wondered how spending more time on math and spelling could make school easier. But Mom insisted that this class would be "good for you." "Good for you"—not exactly the words of wisdom and encouragement I was looking for.

However, at ten years old you don't have a lot of bargaining power on these issues. What choice did I have? If I complained, I would only get in trouble and accused of being lazy or irresponsible for not cooperating. I convinced myself that if I worked hard enough at summer school, maybe teachers would stop yelling and criticizing me. And I might be able to find a page in a book during class without being laughed at by a classmate.

School let out, filling me with mixed emotions. On one hand, I felt a sense of incredible relief as I left Mrs. Owen's fourth-grade class. But, at the same time, I knew class at the clinic would start soon and that I would be doomed to an apparent lifetime of calculating hundreds of math problems and reciting spelling lists several pages long.

Self-doubt, criticism, and negative thoughts raced through my mind, making the knots in my stomach twist tighter and tighter as summer arrived. What if I get the problems wrong? What if Carol yells at me? What if she puts all of the assignments on the chalkboard like Mrs. Owen?

As I thought about these questions, tears formed in my eyes as quickly as I wiped them away. My nervousness was only intensified by the fact that I had to keep my worries inside. No one, including my parents, wanted to listen. Instead of seeing my desperation and fear, Mom assumed that I was being my stubborn self and simply refusing to cooperate. However, I knew I had no choice but to give Carol's summer school a try.

૭ ૭ ૭

> " 'Good for you'—not exactly the words of wisdom and encouragement I was looking for. "

Carol: Planning the Constructions Class

Now that I had committed to a summer class, apprehension gripped me. This was supposed to be a class for children with spatial learning problems. Three other children besides Britt had registered. None of these kids really understood numbers despite having been taught over and over in school and at home. Could this class really make a difference?

At the office, I poured a cup of coffee, took out a new writing pad, and started to plan. Fortunately, we had moved to a larger space that included a room with two eight-foot tables. I reserved the room for the 9 to 12 slot during the first two weeks of July. That meant 12 hours of class time per week since we would not meet on Fridays.

The guiding premise would be discovery. A lab class. Better call it "Constructions." In a "Constructions Class" you do and you make. The children would understand that. Discovery was fun, but that was not the reason to use it as method. Instead it was because discovery let the children integrate their own learning. They could only "discover" what they were ready for—what they had built a base to learn. While the children in the class had similar problems, they certainly were not identical. Each child needed to move at his or her own rate. Besides, discovery allowed leaps of progress when critical pieces of understanding fell into place.

I realized that the concept was really outrageous. Shape perception. Fill in the gaps. Build a base for understanding space. And be sure to teach the language that goes with the concepts. I set about planning a series of activities that would require each child to build a model for shape, dimension, position, and quantity.

Again I wondered how these children had missed out in learning this basic information? Something early in life had to happen. Of course, we knew that Britt's medical condition made visual-spatial activities very hard. Tom was a preemie who had to have supplemental oxygen for a long time. Jeremy had a cast on one leg or the other most of the time until he was three or older, so he was unable to move around. I didn't know if Morgan had any health problems in early childhood, but then, she had the fewest problems now. The common factor for all of them, according to the tests, was that they lacked an accurate mental model for number (e.g., how much is *eight*) and quantity (e.g., how much is a pound). During assessment, Tom estimated that his chair—his eight-pound chair—weighed two hundred pounds. And he was serious.

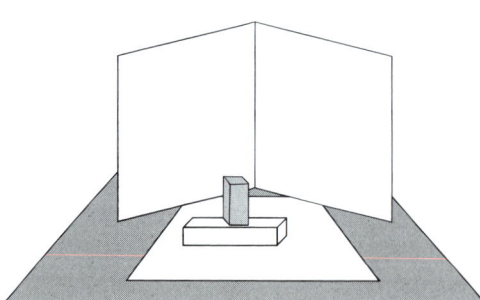

These children needed to learn specific language to describe space. But they could not use descriptive words for position and location unless these words were in their vocabularies.

I decided we would warm up for a few minutes each day with the "block game." Each child would have identical colored blocks and a screen to conceal his or her layout from the others. The rules would be that whoever was "it" described how and where he or she placed his or her blocks. The speaker had to be precise, but so did the listeners if they were to follow the directions. I could imagine the disgust Tom would express if the speaker used vague words that he could not follow. They would learn new descriptive words, such as *parallel*, and become more specific in word choice as the other class members demanded clear language.

I planned every free moment. Constructions class meant constructing, which is the very activity these children had

Carol: Planning the Constructions Class, *continued*

avoided. We needed to soften the process as much as possible. Using one-inch graph paper for the constructions would simplify determining area and also provide guidelines for cutting. They would need a few basic paper skills like cutting and pasting.

After the children made paper boxes that were 2"x 2"x 3" and filled them with wooden one-inch blocks, 12 cubic inches should make more sense. Then they would have the task of figuring out how the math works to arrive at that answer. Discovery was the key. I wondered how close to their predictions their final answers would be.

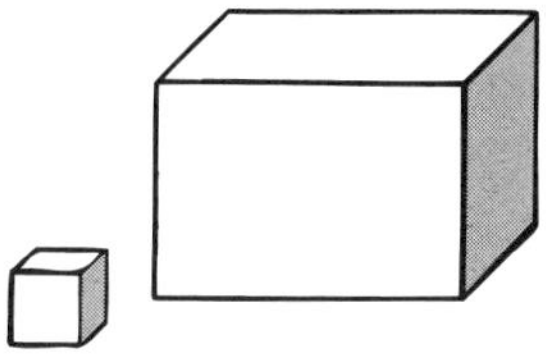

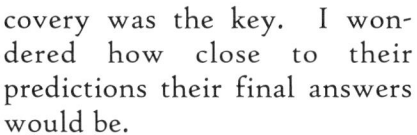

We would use the Learning Window to shape our understanding of basic forms such as boxes and cylinders. I had tried it out with several students and it seemed to be effective. The students would trace the objects below eye level, at eye level, and above eye level. Then they would compare the views. Maybe they would be able to predict viewpoint differences and take other points of view after this experience. They certainly couldn't do it now.

We would also use the window to trace one another and ourselves with the help of a mirror. After all, if we were learning to measure, and our bodies are what most of us use as a base, we needed to sharpen our image of ourselves.

So many small confusions to overcome. I wanted to see Britt's face after she drew around her hand and then walked around to the other side of the window and held her same hand against the outline. I wondered if she would feel as if her thumb had moved.

My list of activities grew, but uncertainty remained. I had not done this before and a host of self-doubts gnawed at me. Such a simple goal—perceive space and number. Maybe these children couldn't learn this way. Maybe the developmental timing was off. Maybe the learning would not transfer to math.

I thought of the many others whose ideas and models had preceded this class: Robert McKim and the work on Visual Thinking at Stanford University, and the Visual Literacy Association members and their visual skills programs. I had read and reread books by professors Rudolph Arnheim and Dondis Dondee on shaping perception. Once again I mulled over what I had seen at the Kingsbury Lab School in Washington, D.C. They required constructions as part of the foundation for overcoming learning disabilities.

My little class was following giants. I wondered if they had ever doubted what they were doing, tormented by the temptation that not doing anything would certainly be much easier than breaking new ground. Somehow I suspected the first step for each of them had been to overcome self-doubts. I shrugged my shoulders and reached for a fresh sheet of paper. The time had come to act on faith. After all, that's how hope is born.

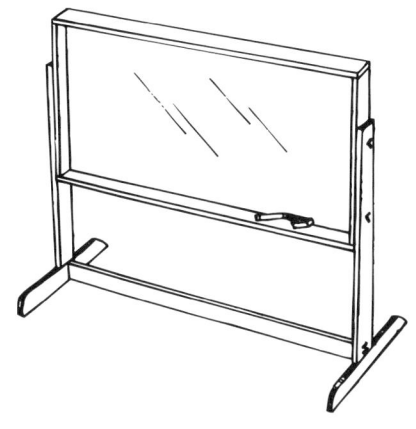

ಬ ಬ ಬ

The Source for Visual-Spatial Disorders

Julie: Did I Waste My Money?

The first day of Britt's summer class arrived warm and clear. I knew Britt wasn't happy about the class. She rode in silence for the 20-minute trip to Carol's office.

"Just let me off at the side door," Britt said, curtly. "I've been here a hundred times."

"Okay," I replied. "I'll meet you right here at noon. Have a good time!" Britt ignored my final words.

Leaving to do errands, I felt both relieved that the class had started and a bit guilty for forcing Britt to do something that she was so unhappy about. I just hoped it would help her with her problems at school. If it did, Britt would be happier in the long run.

I stopped to visit a friend who had taught third grade for years. I told her about Britt's summer class and my doubts about it. She said she had seen many children like Britt who were just slow at learning certain things. "Britt probably just needs more practice," she said. My friend came out of her study with a thick folder full of math practice sheets. "Have Britt do a few of these each day, and she'll be up to speed by the time school starts in the fall. Trust me, I've taught for 20 years. She'll be fine."

I tucked the folder into my bag and wondered if math drills really were the answer.

When I returned to Carol's office, I was braced for a sullen, angry Britt. Instead, a smiling Britt bounded to the car.

"How was it?" I asked.

"Good," Britt replied. "We didn't do any math or spelling. And guess what? Morgan is in my class."

"You mean the 'Morgan' who was our neighbor before we moved to Gig Harbor?" I asked.

"Yes. I haven't seen her for so long. Can she come over to play tomorrow after class?"

"Sure," I said, relieved at Britt's enthusiasm. And then I asked, "Are you sure you didn't do any math?"

"No math," Britt replied. "We played with blocks and folders, and then we talked about them. It was really fun."

I nodded, but said nothing. This didn't sound like a math course. Maybe this was only a warm-up. Surely tomorrow they would do real math.

The next morning, Britt was eating cereal before I came into the kitchen. "Don't you think you better get dressed?" Britt said to me. "I don't want to be late to class."

"It's only 7 o'clock. We have plenty of time," I said.

After class, I picked up Britt and Morgan, who walked to the car arm in arm giggling.

"How was class?" I asked.

"Great," Britt said. "We played with blocks and beads. It was fun."

The girls played until Morgan's mother, Claire, picked her up at 5:00. Claire was as perplexed about the class as I was, but she shrugged and said we would just have to wait and see.

Each morning Britt was enthusiastic about the class. When I picked her up, she was excited about an activity that sounded as if it had nothing to do with math. I was relieved she was happy, but I still worried I had wasted my money on unproven methodology that had done nothing more than provide Britt with an entertaining interlude in her summer.

> **" *This didn't sound like a math course. Maybe this was only a warm-up.* "**

ಬ ಬ ಬ

The Source for Visual-Spatial Disorders — Copyright © 2002 LinguiSystems, Inc.

Britt: Summer School

It was a brilliant July day when I got out of the car on the first day of class. "Have a good time!" Mom said as she waved and pulled away from the curb.

Panicked, afraid, and unprepared for what I perceived as a math and spelling marathon, I made my way through the medical complex to the clinic office. Anxious and trembling, I took a deep breath and opened the door to the clinic waiting room, forcing a big smile and a cheery "Hello," hoping that using a lively voice would ease my nervousness.

Walking into the clinic waiting room, I realized that I was not the only one in the class. Morgan, a long-time friend, ran up to me as I opened the door. Then, for the first time, I began to breathe a sigh of relief and wondered if the summer might redeem itself after all. I had known Morgan almost my entire life, which at that point was about ten years. We began to chatter about school, our annoying brothers, and the class we were about to start.

Two scruffy boys sat across the waiting room glancing nervously around the room and trying to decipher whether Morgan and I were there for the same reason.

Morgan and I ignored them, continuing to banter and giggle intermittently. Our conversation came to a halt as we were summoned to the classroom in the back of the clinic.

Carol greeted the class and started telling us what we were going to be learning over the summer. Carol explained that all of the things she had planned for the class would help us learn about number and space.

"Ah ha," I thought to myself, "here it comes. She is going to make us do math after all."

I shifted in my chair, swinging my legs back and forth to ease the nervousness and gave Morgan a reassuring half grin. To my surprise, class started off with a game. Once we started playing, we put all of our doubts behind us and focused on the task at hand. Morgan started the game off by volunteering to be "it."

"Okay, follow my direction and do everything I say. Just like the game 'Simon Says,'" Morgan instructed as she began arranging the blocks behind the screen. Then she told the rest of us how to assemble our piles of blocks to match hers.

As we moved our blocks around, I asked Morgan questions. Her words helped me see how she had arranged the blocks on her side of the screen.

While piecing the blocks together, I realized that the game was like a puzzle. Only this time the pieces fit, and I did not have to blindly guess how the parts fit together as a whole.

"How did I do?" I asked as Carol lifted the screen revealing Morgan's pile of blocks.

Perplexed, I wondered why Morgan's block arrangement looked so very different from mine. Momentary panic struck as I envisioned having ruined the entire game by confusing

> " *Once we started playing, we put all of our doubts behind us and focused on the task at hand.* "

right and left. But when I ran to view Morgan's side, I was surprised that I had actually been able to talk through the puzzle and learn direction. It was just a matter of perspective.

Once we recovered from the game and settled back into our seats, Carol threw out a bunch of beads on the table. Some beads were separate and some were wired into clusters of 10's or 100's. As we reached to pick up the random beads off the table, we held them in our hands registering the weight and number of beads in each grouping.

The Source for Visual-Spatial Disorders

Britt: Summer School, continued

Carol asked if anyone knew how many beads there were in each set. Everyone shook their heads. Curious, we moved the groups of colored beads between our fingers, counting the clicks as they slid down the wire. We began to notice a pattern. Realizing that ten beads in a row were always the same size and weight, I used a second set of beads to compare. For the larger numbers, like one hundred, the beads were grouped together as a whole block. Rotating the beads with our fingers gave us a feeling for the number represented in each piece. Each bead turned separately as we brushed our palms across the side of the blocks. Soon the quiet room filled with excited noise as we made our discoveries.

"Look at this! It really feels like one hundred," Morgan said as her hand swished across the block. One, ten, and one hundred. Every bead in the block revealed new meaning behind numbers we could not seem to understand.

We moved on to other projects involving cutting and folding paper. We cut, folded, curled, and taped paper into three-dimensional shapes, turning spatial relationships into language we could relate to. We cut circles out of graph paper. Carefully, I used scissors to follow the line I had made with a compass. We folded the circles into cones and set them on the table. The flat, flimsy paper miraculously took shape, instantly turning into Indian tepees from one of the Laura Ingalls Wilder books I had been reading. Morgan and I kept making tepees until we had a whole village, all different shapes and sizes.

Class was over and I bounded out of the clinic, racing down the stairs to meet Mom. Feeling a sense of relief that you only know after being worried for days, I leapt into the car. Mom asked me what I had done in class and I told her that we played blocks and made tepees out of graph paper. I saw the puzzled look on her face and explained that we had fun and that we did not have hundreds of math problems or spelling words to worry about.

"What?" Mom said. "You mean that you aren't practicing math and spelling?"

"No, not right now," I replied and was suddenly brought right back to reality.

This was only the first day, and we still might be stuck with math problems later in the class. The sense of worry and dread that had been with me for so long began to settle in again. But I did not care. I felt as if a huge rock had been lifted out of the pit of my stomach.

"Guess who's in class with me?" I asked. "Morgan. Please, Mom, may I have Morgan over after class tomorrow?" Mom agreed.

As we arrived home, Mom reminded me to sit down before dinner and work on a sheet of math problems before I could play outside with my brother or read. I sat and stared at the thick folder stuffed with sheets of numbers just waiting to be calculated. I wondered if I would ever understand math and wanted desperately to escape. Surely, the characters in the book I was reading would have found an imaginative way to cope with this problem. But, this was reality, and I decided just to try and finish the problems without complaining.

Having made it through the first session, I had different expectations about what we would be doing with Carol and wondered if the next class would be as fun. If nothing else, at least Morgan would be coming to my house after class. Besides, Carol always had kind words for us. And, as long as Carol never embarrassed us by

> *" Every bead in the block revealed new meaning behind numbers we could not seem to understand. "*

The Source for Visual-Spatial Disorders

asking us to do math problems on the board in front of the class or participate in a spelling bee, everything would be okay.

The next morning I opened the door with less trepidation, let out a cheery "Hello," and started to discuss plans for the afternoon with Morgan. "After lunch we should make a fort in the woods. I'm sure that my brother will help," I suggested.

"Wow! You have a place in your backyard for a fort?" Morgan wondered.

"Yeah, and maybe more than one. We could build our own town, just like the one in *Little House on the Prairie*." Our conversation continued as we filed into the classroom. The session was just about to start.

Carol began by reviewing the game from the day before. Several new blocks appeared in the piles she handed out. Now anxious to play our new version of "Simon Says," we set up the manila folders and started arranging our blocks carefully, describing our movements to our partner on the other side of the screen.

"That was fast!" Carol exclaimed. "You are getting really good with the blocks," she commented.

Discussion about the outcome of each game easily flowed into conversations about ways to find shapes in different parts of our environment. Everyone agreed that we could find these shapes in all sorts of different places. With one hand, Carol held up an enlarged black and white picture of a house mounted onto a neon piece of cardboard as her other hand waved a bright red marker.

"Let's say you were standing on the ground right next to the front door looking up. What shape is the roof from this perspective?" Carol asked the boy sitting in the seat on my right as she pointed to the area in question with the marker.

"Hum, I guess it's a triangle—I think that's what I'd call it—maybe," he replied hesitantly.

"Okay, that's good. Yes, the roof of this house is a triangle when you look at it that way." Carol confirmed this fact as she traced the roof with the bright red marker.

"Now, what about the front of the house? What shape is the door?" All four hands went straight into the air volunteering answers.

"I know, that's a square," said the other boy sitting two seats down from me.

"Not quite," said Carol.

"What about a rectangle?" Morgan answered.

"Yes, a rectangle looks a lot like a square, but what is the difference?" Carol asked the class.

"The sides of a rectangle are longer," I quickly replied.

"Great!" Carol said, "That's exactly right. There are four sides in both a square and a rectangle,

> " *Now anxious to play our new version of 'Simon Says,' we set up the manila folders and started arranging our blocks carefully.* "

but a rectangle has two sides that are longer than the other two sides."

Carol illustrated this point by drawing around the door of the house and showing us the longer sides. She kept showing us pictures and pointing out shapes. Finding pictures a little dull, we identified shapes around the room, pointing them out as we walked around, preparing for the next activity.

Then we looked at the beads we had seen the day before. I remembered that Carol asked us to guess the number of beads just like guessing how many

Britt: Summer School, *continued*

jelly beans in a jar. This time we were familiar enough with the denominations represented that we did not have to guess. All we had to do was feel the difference. Then Carol placed a handful of beads on the table.

"How many beads are there on the table?" Carol inquired.

Concentrating, we carefully picked up each block and studied each bead until we were able to come up with a number.

Once we agreed on the number, we recorded it on the graph paper Carol had passed out. I learned how much easier it was to keep track of number columns when they lined up in the appropriate rows. I did not have to worry about the numbers jumping all over the page. I now had a fighting chance to conquer math problems.

Carol did this exercise more than once, giving us a chance to practice. After a while, she had us add and subtract beads to the collection already on the table. Morgan and I worked hard, not always coming up with the right number, but never reluctant to try again. Carol told us this was math. We were learning to count, add, and subtract using these beads. We learned how to represent the problem 300 - 21 as well as 300 + 21 all using the bead method.

We finished class by cutting and taping paper into different shapes. We made brightly hued pinwheels that spun when we blew on them. Morgan and I ran out of the classroom, holding our pinwheels as they twirled in front of us.

> **" I now had a fighting chance to conquer math problems. "**

෴ ෴ ෴

Carol: Beads and Blocks

The giggling in the waiting room told me that the two girls had discovered one another on this first day of class. The boys sat at the other end of the room warily eyeing the rest of the room. Sighing, I looked over the stacks of one-inch graph paper, scissors, blocks, manila folders, tissue, tape, glue, large sheets of paper, and stacks of pictures. I was ready, as ready as possible.

The block game was a hit. That and the realization that they would work together in teams and that we were not doing sheets of math problems brought smiles from all four.

Now they were trying to stump one another with the game that I had modeled first with Morgan, then with Britt who volunteered. Each player had a folder and identical sets of blocks, a red square block and a blue rectangular one. Hidden behind the folder, the lead player described how he or she placed the blocks and the others tried to duplicate it.

"I am laying the blue block flat on the table with the wide side down," said Britt. Behind the folders, the others followed the instruction.

"I am putting the red block on the right of the blue block," she added. Next, they raised their folders to compare arrangements. Tom had mixed right and left. Britt's and Jeremy's were almost alike. Morgan had placed the red block several inches away from the blue block. Now they were arguing.

"You didn't say they were touching."

"Yours is on top; you said on the right."

"Well, I meant on top and on the right."

The game was working. They were using words such as *beside*, *beneath*, *right*, and *left* in this first attempt to describe spatial relationships. And they were shaping one another's choice of words and sentences for greater clarity.

In the days ahead, we added blocks of various shapes and sizes, along with words like *horizontal*, *vertical*, and *right angle*. As they put their blocks and folders into the basket Morgan was passing, Tom announced, "I get to be 'it' tomorrow. I'll stump everyone!"

After putting away the blocks, we moved on to "basic paper skills." I explained that making things required the ability to accurately fold, crease, curl, and cut paper just as baseball required the players to know how to catch, throw, and bat.

Britt's small hands moved efficiently with the paper. Tom tried over and over to fold his graph paper without creasing it. Each time the edge wavered unevenly. Jeremy watched, but did not do any of the activities until I put a pair of scissors in his hands and said, "Go for it."

By now the others were drawing circles with a compass, cutting them out, cutting a line to the center of the circle, and overlapping the edges to make a cone.

"Tepees," said Britt, as I helped her tape the overlapped paper into position. Jeremy picked up one of the pages of circles Morgan had drawn and started to cut. The scissors thrust upward and bent the paper so he couldn't see the line. He changed position several times after taking small snips that veered to the right or left. His circle had two flat sides and a jagged border.

He threw the crumpled paper into the wastebasket with a quick glance around to see whether any of the others noticed his crude cutting.

The Source for Visual-Spatial Disorders

Carol: Beads and Blocks, continued

"Jeremy," I said, "these scissors may be sharper," as I handed him a smaller pair hoping they would be easier to control.

He sighed and left them on the table until I urged him to "cut one circle anyway."

By this time, the others had cones (tepees) of all sizes and shapes arranged on the table. Hiding behind the pile of paper scraps on the table, Jeremy tried again. Very slowly, he maneuvered his scissors and then quietly slid the finished product through the paper scraps to me.

I caught his eye. "Good job—that's hanging in there," I said.

The circle looked chewed, but at least it was recognizable as a circle. Jeremy's fine motor control was a problem—a big problem. His school report described him as lazy. What an effort writing must be!

I slowly spilled the large tray of beads on the table so that they were within everyone's reach. The beads were loose or wired together into sticks, squares, and cubes. I asked the class to work together to decide how the beads were grouped.

Britt, Morgan, and Jeremy counted and compared. Tom argued. Eventually they all decided that the sticks of beads held ten beads and the flat squares were made of ten sticks. Tom argued for 88 beads in the square, but after counting them many times, the group agreed on a hundred.

The large cube of beads was a puzzle. No one knew or could figure out how many beads were in the cube although Tom said it wasn't a regular number. The others looked at him quizzically. I did not reply.

Next, we each marked the right column on one-inch graph paper with a wide yellow marker. No mistaking that column. These were our "math sheets." Only one number could be written in a square and the first square to get a number was the one marked in yellow.

All these kids had problems with directional confusion that showed up in many ways. For instance, "13" might be written "31" and, in turn, read as either number. Subtraction was very confusing in that sometimes they took the bottom number from the top and sometimes the other way around.

I tossed three beads on the table and asked the teams to record the number correctly on their sheets. They looked at one another as if this was just too insulting, but they put a 3 in the right-hand column. Next, I retrieved the loose beads and dropped three sticks of beads on the table.

Tom tried to write 310 in the right hand square, but Jeremy stopped him with the reminder, "only one number in a square."

Morgan knew that the three sticks were 30 beads and wrote 0 in the right-hand square and 3 in the next square. In a few minutes, the entire group caught on to the pattern. They could translate any bead combination into numbers and then write them correctly in the columns.

Soon the teams began a game. The girls thought up a number and the boys had to make it in beads and write it. Then the groups switched roles. One of the other teachers stuck her head out of her office and "shushed us" when all four voices began defending their answers at once. Discovery was working!

After other activities, we ended the class by making another three-dimensional shape. The function of this activity was to enlarge their "spatial" vocabulary to include fractions.

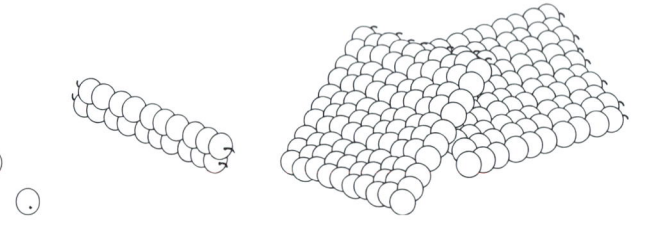

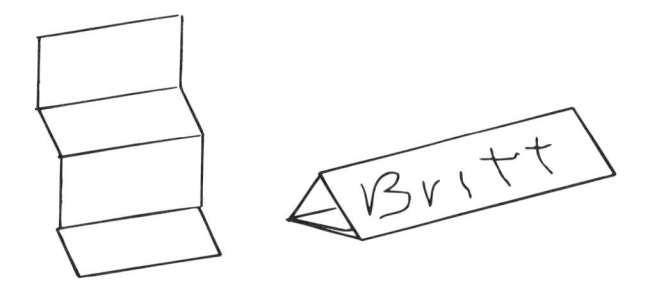

Carol: Beads and Blocks, continued

Each student folded a 3 x 5 card in half and then in half again. We discussed how many divisions were in the card and they easily answered such questions as, "Show me ½. Show me 4/4." Each student then wrote his or her name on one panel of the card, overlapped the end panels, and paper clipped them together. They would use this nameplate for each subsequent session. A task for the next day would be to figure out an addition or subtraction problem in fractions using the nameplate.

As they bounced out through the waiting room, Britt and Morgan were chatting non-stop. Even Tom was less gloomy. I heard him say to Jeremy, "Well today wasn't too bad. Tomorrow will probably be dumb, though."

The door swung shut behind them. Soon we would start working with the Learning Window. And do much more with dimension.

ಐ ಐ ಐ

Carol: A Window Joins More Beads and Blocks

As soon as I entered the room, Morgan blurted out, "A thousand—it has to be a thousand!" She could not wait to predict how many beads were in the heavy cube they had all handled the previous day.

"Do you all agree?" I asked.

"No way!" said Tom emphatically. "It's maybe—maybe a million," he added as he grabbed the block of beads from Morgan's hand.

Together we counted 100 per layer and piled the layers until there were 10. Morgan moved her finger up the stack as she counted, "One hundred, two hundred, three hundred..." She smiled and said emphatically, "One thousand!" as she reached the top.

Tom was appalled, "But what's a million?" he pleaded.

"A thousand thousands," I replied. He was stunned.

"That's like every one of these beads turned into a thousand!" he announced, pretending to drop the heavy block he cradled in his hands.

We moved into the classroom and they began with the warm-up exercises with the graph paper we had learned to use the first day. They could easily write numbers of any size. In minutes, they were adjusting numbers up and down, and describing each action. Soon we were writing these as addition and subtraction problems. The children had bridged from handling beads to more traditional-appearing mathematics. More bridges would be necessary, but this was a successful beginning.

Each of these children must have used "manipulatives" like the beads in kindergarten or first grade. Yet this approach had not worked for them before this. Now manipulatives were working. I considered reasons why this might be so. They may not have been developmentally ready before. Maybe it helped to have them predict answers. This step seemed to give them a comparison base. Then, too, the graph paper for recording answers helped limit directional confusion. We talked about how each square served a different purpose. Jeremy, solemn as usual, had observed that the line between the squares was a wall.

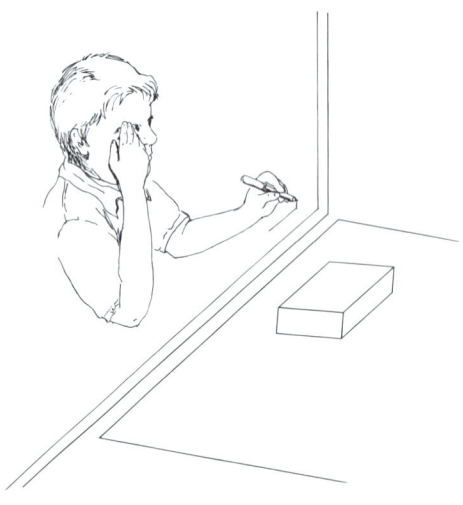

Touch was important for this lively bunch and they all loved handling the beads. I had never noticed the humming sound possible from rubbing the blocks. They literally "felt" the different amounts.

Probably the most important part was "talking the system." The students did the usual activities—constructed numbers, wrote numbers, solved problems—but describing each process to a teammate seemed important. The teammate's complaints if the directions were not clear had more impact than anything I might ever say.

Curiosity was mixed with excitement when I brought out the new Plexiglas Learning Window, pens, and damp sponges for erasing. Tom couldn't "read" most pictures, which meant he missed a lot of information that was obvious to others. All four of them had difficulty with distance cues and with picturing objects in different positions. Of course, this is a developmental skill, but they were old enough to understand and make pictures.

I showed them how to trace objects starting with boxes. They loved tracing on the window. It was probably the

Carol: A Window Joins More Beads and Blocks, *continued*

first time they could "draw" without being embarrassed. The box tracing was easy for all of them. Next we traced a coffee mug, carefully holding it at eye level.

Mug in hand, I then asked the group, "What does a coffee mug look like if you are looking down on it from a window in the ceiling?"

"The ceiling!" exclaimed Tom, "There's no window in the ceiling."

Jeremy was looking down, trying to avoid my eyes because he was afraid I was going to ask him to draw. But even he smiled at Tom's ever-present literalness.

I asked Tom to stand up so he was above the mug on the table as he traced it. He sat down and looked at his drawing, then at the mug in his hand.

"Hey, my drawing looks like that cup looks. Can I take my drawing home?" he asked.

I slid the Plexiglas out of its frame and took it to the copy machine to copy for everyone. Tom had never before realized how point-of-view changes the appearance of an object. We drew boxes, cans, and other objects at various perspectives. Soon they were locating pictures in magazines for each of the perspectives.

Another bridge had been built. Spatial position was linked to appearance. Before long, each of them would be able to predict how objects looked from various angles. This meant

> **" *Another bridge had been built. Spatial position was linked to appearance.* "**

that they could also create images that flexibly shifted to suit changes in position. Flexibility in the use of imaged objects was growing. Precise images were tremendous support for calculation and reading comprehension. I smiled as I put the beads back in the cupboard. All four students were building the fundamental skill necessary to solve story problems in math.

ಙ ಙ ಙ

The Source for Visual-Spatial Disorders

Britt: Discovering Shapes

Happiness was now coming into class and learning. We tried anything that Carol wanted us to try. Perched on top of the Formica table was a glass window. Next to the window were dry-erase pens. Carol told us this was called a Learning Window.

"Pick up a pen and trace your hands with any color ink you want," Carol instructed. We traced our hands and ran around to the other side, eager to see how they looked.

"Wow, my hand has been turned around. It actually looks like I'm waving to someone," I exclaimed as I placed my palm against the outline of my hand.

"That's right," said Carol. "Now, what I want you to do is try tracing each other's hands and arms from the other side of the window like this."

Carol placed her hand on the glass, and I began carefully to trace her hand with the dry-erase pen from the other side of the glass.

"I'm pretty good at this," I said as I quickly made my way around the outline of Carol's hand.

"Yes, and everyone gets a chance to try drawing on this window," Carol commented as she stood up and divided us into our usual teams. We started off taking turns tracing simple objects such as our hands and arms. After we had a chance to do that, we moved on to more complicated objects.

"Now, we are going to look at this block and trace it from several different angles," Carol said. She held up the block on one side of the window while Morgan traced the block from that perspective. Carol moved the block six times and each time the block was traced to show the changed perspective.

Carol pointed to one of the drawings, "This is what the block would look like if you were a mole."

"Or a rabbit," I added as Carol pointed to the top tracing on the window.

"This is what the block looks like if you are a person and it is in front of you," Carol said as she indicated the middle drawing. "And this is how it would look if you were a bird."

"Or in an airplane," I interjected proudly as Carol pointed to the last tracing toward the bottom of the window.

Following that demonstration, we broke into groups and began taking turns drawing the cube as one partner held up the block for the other to trace. Once we had mastered drawing the cube, we moved on to coffee mugs, books, and folders all from different points of view.

After using the Learning Window, we had a chance to make paper airplanes. Carol gave us instructions, and we followed them.

> **"** *Mom kept handing me scores of the math sheets procured from a seemingly helpful teacher friend.* **"**

Every afternoon in the car on the way home, Mom asked me if we had practiced math and spelling. My answer was always the same, "No." Still insistent that drills would benefit me in the long run, Mom kept handing me scores of the math sheets procured from a seemingly helpful teacher friend. Despite all of the work at the clinic, I continued to spend the late afternoons and evenings crouched over the math problems at the kitchen counter. While Mom started dinner, I did math. 834 + 124 might come out 876, 985 or, if I happened to get lucky, the correct answer, 958.

Even after an entire week, I still found myself looking forward to going to class.

"Today," Carol said, "we are going to learn about measurement. How many of you can show me where an inch is on

The Source for Visual-Spatial Disorders

Britt: Discovering Shapes, *continued*

your finger?" We each made a guess, but all of us were off by at least two inches or more.

"An inch is bigger than your finger isn't it?" I heard one of the other students ask.

Carol proceeded to pull out plastic cups, a bottle of ink, and a stack of wooden rulers from a cupboard at the back of the room. She passed out a cup and ruler to each student and asked us to hold our ruler in our cup as she came around the room pouring ink into them. When the ink reached an inch on the ruler, she stopped.

"That's an inch," Carol said.

Curious, we each pulled our stained ruler out of the ink to see how much of the ruler had changed color.

"The next part will be fun!" Carol told us. "I want you to dip your finger into the ink."

The thought of making a mess on purpose caught us off guard. However, I liked the idea of slopping ink around with my fingers and followed instructions. As I pulled my dripping finger out of the cup, my eyes opened wide with amazement and realization. Amazing! I now had a sense of how big things were in relationship to myself. With this mark on my finger, I could estimate the length of my arm, chair, and even the table. After studying our fingers and measuring them against our rulers, we went around the room measuring objects.

"That lamp shade is fourteen inches!" I yelled, counting out the inches as I moved my finger up the shade.

"Hey, this coffee mug here is four inches tall, and the pen next to it is six inches," I shouted enthusiastically.

"Wait, look at this! You won't believe that the table leg is twenty-seven inches long!" Morgan exclaimed.

These little bits of revelation went on for the rest of the morning. The excitement of discovery filled the room and gave us confidence to continue learning and exploring new methods for conquering the world around us.

When I came out of class that day, Mom asked me what we had done. I proudly displayed my stained finger.

My brother John turned around in his seat to look back and carefully studied the mark on my finger. After a pause, I heard him say, "I think your blue finger looks stupid."

I ignored my brother and looked over at Mom. Even though she did not say anything, I could tell by the look on her face that she agreed.

Finally, Mom asked, "Why on earth do you have ink on your finger?"

"That's an inch," I explained.

"Yes, that may be true, but *why* do you have an inch marked on your finger?" she asked again.

"We learned how to measure using the ink marks on our finger," I replied.

Again her eyebrow slowly raised itself, skeptical of the idea of spending an entire day measuring things.

Before class had ended, Carol had passed out a list of household items for us to measure. I pulled out the list and showed Mom what we were supposed to do for the next day's session. Once home, I scoured the house for the items we were supposed to measure:

- Length and height of bed
- Length of bedroom
- Length and width of couch
- Width of front door
- Length from bottom of door to door knob
- Height of favorite stuffed animal
- Length of kitchen or dining room
- Height from floor to top of kitchen counter

I went around the house finding things to measure and record. My brother followed me around wondering why I

Britt: Discovering Shapes, continued

was measuring everything in sight. I was so excited about learning this new technique that I finished the entire assignment before dinner and hounded Mom for things to measure while she was cooking.

"Please, let me measure that bowl before you use it for the salad," I begged. "Oh, and that long wooden spoon too," I quickly added.

"Okay, and here is a box of rice to add to your list," Mom said as she handed over the requested items.

I carefully set the items I was about to measure on top of the kitchen counter and hopped up on the tall bar stool. I pulled out my ruler from my back pocket and placed it over the top of the bowl. Exactly twelve inches across, I said to myself. Then I used my finger to check the measurement. Satisfied that I was correct, I moved on to the spoon and rice box repeating the procedure with the same precision.

I jumped down from the stool and went to find a few more items to measure. I walked out the front door and came across my cat, Muffin. Hmmm, I thought. I wonder if Muffin would mind if I measured her.

I picked her up and held the ruler to measure the length of her body. Exactly 20 inches head to tail. Done with the assignment, I made my way back to the kitchen and begged Mom to release me from working on math problems since it was already late and close to dinnertime.

"Oh, all right, but only for tonight," she said reluctantly, conceding one evening of study.

I could not wait to get to class the next day and show Carol how I had completed my assignment. Accomplishing this task even just three days earlier would have been almost impossible. Eager to continue with the assignment, I bounded breathless into the classroom with my paper in hand.

"Look at this, Carol!" I blurted out even before she could call the class to order.

I shoved the paper with all of the measuring I had done the previous afternoon in her direction across the table.

"That's great!" she exclaimed, catching the paper before it slid onto the floor. "How did everyone else do with the assignment?"

She proceeded to read the papers out loud and discuss what we had measured and discovered using our newfound skill. Next we used our understanding of measurement to draw pictures to scale.

Carol walked around the room hanging up four large pieces of white parchment paper on the walls. We all wondered what we were going to do with such big paper.

Seeing our questioning looks, Carol said, "We are going to learn how to trace our shadows. We have already practiced tracing our hands and faces with the Window; now I want you to try using paper and pencil."

She showed us how to position ourselves in the light so that it would cast a shadow on the paper in back of us. Anxious to try this assignment, Morgan wanted to trace my picture first, so I let her. I sat perfectly straight with my eyes focused in front of me and waited until she had completed her tracing. Eager to see the portrait, I jumped up from the chair to see her drawing.

"Incredible, it's nearly life-sized! And, that really is the way my face looks from the side! I hope that I can draw as well as you, Morgan," I said.

Morgan joined me in my scrutiny of the tracing and

> " *I was so excited about learning this new technique that I finished the entire assignment before dinner.* "

The Source for Visual-Spatial Disorders

agreed that it turned out very well, but admitted that she was confused about the relationship of the features to the profile.

"How are we supposed to draw the eyes? If we draw ovals where the eyes are supposed to fit, the picture won't look right," I mused. Uncertain, Morgan and I raised our hands to ask for help.

"That is a good question," Carol remarked. "The eyes should be drawn from a side view just like the portrait. They look like cones from the side." Carol demonstrated by turning her head to the side pointing to her eye.

"You can figure out how to line up the eye correctly by using the top of the ear and the top of the nose as a reference point. Why don't you draw in the eyes and see how it looks?"

After we placed the eyes in the right spot, we were done.

"My turn!" I exclaimed as I raced across the room and grabbed a pencil.

I began slowly and awkwardly making my way around the outline of Morgan's face with the pencil. At first the line was shaky, meandering a little off the path first to the right and then to the left. Discouraged, my hand began to tremble a little. I took a deep breath and decided to retrace some of the drawing. I erased the areas that appeared too far off the mark

> " *As I gained confidence, the line became stronger and smoother, flowing easily around the profile.* "

and retraced the missing lines. I started the drawing with Morgan's hair trying to trace every bump and curve as they appeared on the shadow. Then I added her nose, cheek, lips, chin, and shoulder, completing the picture.

As I gained confidence, the line became stronger and smoother, flowing easily around the profile. Finally the pencil took on a mind of its own and effortlessly made its way to the drawing's finish mark. Finding the patience to complete the task proved the most difficult part of this project; however, I was determined to prove that I could learn to draw.

"There, it's finished," I announced.

Morgan's smile told me she was pleased with the result.

ಬಿ ಬಿ ಬಿ

Julie: Learning About Learning

Much to my amazement, Britt had remained enthusiastic throughout the class. It was as if I had my old Britt back—happy, positive, and confident. I was happy about that, but I continued to wonder if I had wasted my money.

I understood Carol's assessment of Britt's learning problems, but I could not see how beads and blocks, blue fingers, and measuring the cat could help Britt with spelling, math, or copying from the board.

At the end of the second week, Carol invited the parents to the clinic to see what the children had been learning. The parents greeted each other and gazed at the class materials laid out on the tables—beads, blocks, measuring apparatus, ink, paper, rulers, and marking pens. I looked at everything and then I had to ask, "What does this have to do with math?"

Patiently, Carol explained that Britt and the other children, though all of their problems were different, had no foundation for understanding math concepts. This class gave them the foundation. When I left that day, I was struck by how little Britt had known.

Though I had listened, the words only made partial sense. Still puzzled, I made a date to have coffee with Carol the next week. I wanted to understand so I could discover some ways to extend the learning at home. I was finally ready to see that for Britt, the blue finger had more to do with learning math than did another folder of math drills.

Putting a cup of coffee in front of me, Carol said, "You know you have a very special little girl. And how's her cat with the eight-inch tail? Maybe you didn't know, but I asked them to measure many objects and a few of their own choosing. Britt chose her cat. You have a very well-documented cat."

Julie looked thoughtful. "I wondered why Britt was measuring her cat. It seemed—well—a rather different math assignment. I'd love to hear

> **" When I left that day, I was struck by how little Britt had known. "**

about what you did with the class," I said. "Britt just loved it, though it wasn't quite what I expected."

"I enjoyed working with her," Carol said. "She's so willing. And she did learn to measure. More than learning the measurements, she learned to 'feel' how much space an inch occupied, or a foot, or two yards. Once she knew these measures, she could estimate quantities and compare amounts. She could then build relationships."

"The blue finger?" I asked.

Carol responded, "Britt needed to tie the measurements to herself—to her own body. The ink allowed her to become a personal ruler. She also had a wooden ruler for a reference. She measured and cross-checked and estimated and compared until she had internalized *inchness* and *footness*. From this base, she could 'know' the measures commonly used for lineal distances."

Puzzled, I asked, "Why didn't she learn this in school or at home? We use measurement all the time."

"Britt certainly knew the labels," Carol said, "but she couldn't feel the meaning. That is very common for people with spatial problems. Tom thought his chair might weigh 200 pounds. He did not know what one pound was and he did not know how much 200 was, so his estimate sounded absurd."

Carol paused a moment to sip her coffee and went on, "The key to developing this 'knowing' in students with spatial problems is in helping them make the link to their own bodies. After all, the body is always our key measuring device. Even the

The Source for Visual-Spatial Disorders

Julie: Learning About Learning, continued

labels such as *foot* suggest the primary body link. As young children, we learn whether we can crawl under the table without bumping our head. Much of exploration is measuring our environment against our body and our body against our world. We develop a comparative base. Of course, labels like *inch* come later."

"I still don't understand how what you're doing with the sticks, for instance, will help Britt with math," I said.

Carol sighed, "I knew you were going to ask that. It's not that I don't want to tell you. It's just so experimental, and it's so hard to explain. But I do want to try to explain because I think it's so important for Britt, no, not important, essential. It is for anyone with a spatial disability."

I smiled, encouraging Carol to go on.

"What we're doing is building a whole new basis for Britt's spatial understanding. But *new* is probably the wrong word because Britt had such a limited understanding of the spatial world to begin with. She has such a strong verbal base so we're trying to use words to help her build, and understand, the world around her. Language seems to be Britt's key to understanding spatial concepts. But her strong language alone is not enough. She needs to make or build or draw the foundation—the objects she then describes in language. Her memories have to build on real experiences such as an inch of blue at the end of her finger. Now she can feel what an inch is and her word for it can transfer to academic subjects."

"Like math?" I asked.

"Yes. And like thousands of other spatial tasks," Carol said. "Britt will not be able to do math or any other spatial task

> *" Does that mean all of those math and spelling drills might never work for Britt? "*

until she can understand the underlying concepts which are spatially based. These things that we've been doing that look like crafts have really been a way for Britt to build a three-dimensional world while she puts it in her own language. She's very bright so she catches on right away, but she must take this huge step backward before she can go on to the concepts that math is built on."

Sensing my next question, Carol went on, "Just as Britt didn't know what an inch was until we put her finger into the ink, she didn't understand a square or a circle until we made them with yarn and straws. What she learned in this class is incredible. She learned to build a base for images."

Again I was amazed at what Britt hadn't known, and I was amazed at what I hadn't known about Britt. "Okay, does that mean all of those math and spelling drills might never work for Britt?" I asked.

"I don't think they will ever help Britt," Carol replied. "She can learn it, but she'll have to take a different route to learning it. We know now that language is a key to Britt's understanding. We, you, her teachers—all of us need to find ways to help Britt use language to translate spatial tasks and concepts."

I finally began to see how the assessment and the class fit together. Both provided the clues to help us find ways to understand Britt's learning disability and to find ways to help her work around it. Finally, it was clear that the "class" had not been a waste of money. Perhaps at home, I could be a teacher of math in equally creative and three-dimensional ways. I would have to think about how I could help.

ಹ ಹ ಹ

Britt: Ms. Jenkins' Class

Laughter and giggles rose from the bus as it wound around the evergreen-lined streets toward school.

As the bus pulled up to the front of the school, I realized that this would be the last year before junior high. Hoping that school would be different this year, I threw my book bag over my shoulder, joining the horde of pushy students eager to get off the crowded bus.

Ms. Jenkins, my fifth-grade teacher, greeted the students with a smile as they entered the classroom and found their seats. Just by looking at her cheerful face, I could tell this year was going to be different. Gathering us into a circle, we went around the room introducing ourselves and giving our favorite color and song. Ms. Jenkins told us that she loved to sing and that she was even acting in a play produced by the local theater. I knew immediately that I was going to like this class and that Ms. Jenkins would make school fun.

When we discussed the assignments for the year, I knew that Carol's class had helped. Suddenly, the thought of attempting math appeared less daunting, even manageable. Ms. Jenkins let us complete math assignments at our own pace. She set up a chart at the front of the room to keep track of our progress. This chart worried me at first. However, as the year started and we began to complete the assignments, I realized that the first part of the math book was review. Breathing a sigh of relief, I understood that I had a second chance. Armed with pages and pages of graph paper, I made my way through the maze of math problems and numbers. I remember fractions and long division posed the most problems that year.

> **" Breathing a sigh of relief, I understood that I had a second chance. "**

Hesitant to ask for help after my experience with Mrs. Owen, I was amazed at the patience Ms. Jenkins showed when I kept returning to ask questions. I worked the problems over and over until I was ready to move on to the next chapter. Since we could move at our own pace, the pressure to keep up with the math lessons was taken away and I could concentrate on learning.

Ms. Jenkins broke up our lessons with story time and singing. Every afternoon she would read aloud to the class. I reveled in the stories and eagerly waited for the next installment. Before she continued the story, Ms. Jenkins always had someone recap what had happened in the story, allowing us a personal interest in reading. Sometimes we sang songs, one of Ms. Jenkins's favorite things to do. She played her guitar and we all sang along.

In fifth grade, we were introduced to geography and maps. In the back of the classroom, maps of the United States hung with ridges representing mountains and valleys. We spent weeks learning the states and capitals as well as how to read map keys. Although this was difficult for several reasons, we were always in groups and could talk through lessons. Directions such as *east*, *west*, *north*, and *south* confused most of us in the class and measuring distance was a nightmare. Fortunately, Carol had taught me to use my fingers as a way to measure distance, so I had a foundation or a reference point to complete the assignment.

After all of the trouble in fourth grade, I finally had a second chance.

ೞ ೞ ೞ

Julie: Someone in the Kitchen with Mom

With the approach of the new school year, my mood swung from apprehensive to hopeful and back again. The constructions class had been good for Britt, and I was hopeful that many of the problems she had had in fourth grade would be behind her. At the same time, I knew that her learning disability would not go away, even if we could find ways around many of her problems.

On the first day of school, I left work early so I could be home when Britt and John got there. As I took milk from the refrigerator, I could hear Britt's excited footsteps on the walk even before she burst through the back door.

"I'm in Ms. Jenkins' class," she exclaimed. "And she is really nice. She read a poem to us and we got to sing. Ms. Jenkins has a really good voice and she's in a choir. Can we go to the store? I have a list of school supplies I need." Britt stopped for breath.

"Ms. Jenkins sounds wonderful," I said. "I'm really glad you're in her class."

"What's for dinner and when can we go to the store? John needs stuff too, don't you John?" Britt said.

John, who had a mouth full of cookies, nodded and produced a photocopied sheet from his book bag.

"Okay, let's go to the store now before dinner. Bring your lists," I said. Britt and John were already out the door, lists in hand. Britt's enthusiasm

> " *I didn't want to prejudice the teacher against Britt if there was no need to.* "

was contagious, and we drove with the car windows rolled down and music from the radio blaring.

I had been thinking about scheduling an appointment with Ms. Jenkins to talk about Britt's problems, but I was reluctant to burst the bubble. Maybe if I ignored the problem, there wouldn't be one. I didn't want to prejudice the teacher against Britt if there was no need to.

But by the end of the third week, Britt's class had started fractions and the teacher gave weekly spelling tests. Now I had to schedule an appointment to talk to Ms. Jenkins about the work Carol had done with Britt and to enlist her help with Britt in the year ahead.

As I walked down the hall of Britt's elementary school, I noticed an open door with light streaming into the hall. This was Ms. Jenkins' classroom, now empty of children but filled with afternoon sunshine.

Ms. Jenkins stood up when she saw me and made her way to the door through the uneven rows of student desks. She was tall, slender, looked young, and she had a very large smile. She shook my hand and invited me to have a seat at one of the tables. Ms. Jenkins was positive and understanding. Yes, she knew about learning disabilities and she had seen Britt's file. I explained the constructions class and what we had learned about Britt's disability and about her abilities. Ms. Jenkins said that she wanted to make learning enjoyable for all of the children in her class, and she was ready and willing to help. Yes, she would enjoy meeting Carol.

I had a new bounce in my step as I walked to the parking lot. Maybe fifth grade would be a happier place for Britt. I had already thrown out all of those miserable worksheets; now I needed to find different ways, three-dimensional ways, and language to help Britt learn math. Remembering the measuring exercises Britt had done as part of her summer class, I decided that I could help Britt with her fractions in a very positive way.

The next Saturday morning, I interrupted Britt's cartoon

Julie: Someone in the Kitchen with Mom, *continued*

watching. "Hey, Britt," I said, "want to help me make blackberry pancakes?"

"Sure. Now?" Britt asked. Still in pajamas, she came into the kitchen where I had laid out the various measuring cups and spoons we were going to need. Without ever mentioning the dreaded word *fractions*, I filled the measuring cup with flour and then let Britt measure out three one-third cups. I then asked her to pour them back into the one-cup measure and tell me what had happened. In a concrete way, she had learned that 3/3 is the same as one cup. We did the same thing with the half and quarter cups. I told her we were practicing for the pancakes. I explained and Britt measured. Then Britt recited the process to me. By the time we finished, I could tell that she had a better understanding of fractions at a level she could not reach when she only had the textbook to rely on.

We moved to the recipe for pancakes. Britt read, measured, and stirred until she was ready to cook them. The pancakes were delicious and we agreed that we could have eaten more.

Over the next week, we filled the cookie jar and the freezer with brownies, chocolate chip cookies, and Snickerdoodles. A few weeks later, Britt asked if she could make pancakes again. "Let's make more this time, so we have lots of pancakes," Britt suggested.

"Good idea," I replied. "Let me show you how to double the recipe." I explained how to measure twice as much sugar, flour, etc. Then I showed her how to double the recipe. She did the math problem, talked through the solution, wrote out the recipe, and followed it.

As time passed, we tripled and quadrupled recipes, not always with the best results, but always with Britt doing and understanding the math.

When we started, the teaching in the kitchen was sometimes burdensome, but as Britt progressed, she could be a real help. More importantly, the active learning and conversations about the measuring and cooking were helping to build her confidence in ways that doing math drills could not.

The fractions in the kitchen had been so successful that I began to look for other ways to help Britt rely on her strengths. Britt had an excellent sense of what kinds of things went together, so I let her have a major part in the weekly menu planning. Of course, she also had to help with the cooking and shopping, and I taught her how to make grocery lists and to grocery shop, comparing quality, quantity, and prices.

I learned that when I gave Britt tasks to do, I needed to make a list of those tasks and go over them orally with her. I also learned that Britt might never be able to spell *separate* correctly, and it would save her great frustration if I would correct the spelling on her papers before she turned them in.

We also bought Britt an electronic speller that she could keep in her desk at school since looking up words was time consuming and frustrating. While the new strategies didn't solve all of Britt's problems, they certainly helped, and they gave her practical skills and confidence far beyond what she was learning from the math books.

> **"** Over the next week, we filled the cookie jar and the freezer with brownies, chocolate chip cookies, and Snickerdoodles. **"**

ಬ ಬ ಬ

Britt: Pancakes, Cookies, and Pies

I thought Mom wanted me to know how to cook. I only realized much later that she really wanted me to learn my fractions. I thought she just wanted us to have a lot of cookies or pancake batter when she asked me to double the recipes, but she actually wanted me to manipulate the numbers. I became the cookie queen in our house, making lots of molasses, spritz, chocolate chip, and Snickerdoodles. Mom's cookbook still wears stains from years of use bearing witness to this fact. Well worn, the book falls open naturally to the baking sections and favorite recipes I still use today.

Using flour and sugar, Mom showed me how four quarter cups fit into a single cup, allowing me to discover how four parts made up the entire whole cup. I did the same with three third cup measures and with two half cups. We also had a two-cup measure and I learned that it would take eight quarter cups to fill it up; it would also take two cups of flour to make a batch of English toffee cookies.

We made cookies and pancakes. Mom pointed out that a single recipe didn't make many at all. "Why don't we triple it?" she said, handing me a piece of paper so I could do the math before I started. She showed on paper how to triple the flour and then she took me to the flour bin and had me measure it out. I then tripled the oil, sugar, and buttermilk. This is still my favorite recipe for pancakes.

Next I learned to make spaghetti. I learned decimals in terms of pounds of hamburger and ounces of tomatoes. We

> *" I thought she just wanted us to have a lot of cookies or pancake batter when she asked me to double the recipes. "*

translated decimals into fractions and back again, adjusting the amount of meat and tomato according to the recipe. To cook the pasta, we had directions that said, "Use a three-quart pan." Mom showed me how to reduce the quarts to cups and then measure the appropriate cups of water.

Then we graduated to planning menus. This was important in learning how to time food preparation, how to follow complicated instructions, and thinking about proportion (e.g., amount of salad or casserole you need for five people versus two people).

I also learned how to shop at the supermarket and compare prices by looking at labels. How much is one brand of green beans compared to another brand of green beans? How many apples is a pound? We placed the apples in a plastic bag and I could feel how heavy they were. Then I put them on the scale and made the connection between the actual weight and the numbers on the scale. We always examined the recipe and made a detailed list before we went to the grocery store so that we knew exactly what we needed. We also discussed how many people would be at dinner so that we'd have enough food.

I cooked a couple of times a week and eventually had a more basic understanding, not only of fractions, but also of all kinds of measurements. I learned the vocabulary of cookbooks, and I can follow almost any recipe. My friends tell me today that I'm an excellent cook.

౪ ౪ ౪

Britt: Things I Learned in Middle School

At first I was excited when I thought of changing schools and moving on to new challenges. Then I realized that I would have to find my way around a completely unfamiliar place, and I started to worry as I sat on the bus during the 45-minute ride to school. Looking out the window as the bus made its way through downtown Gig Harbor, I wondered if my teachers would like me. I wondered if the kids at school would tease me. Most importantly, I wondered if middle school would be too much for me and whether the teachers would give assignments that I couldn't handle without extra help. The bus struggled up the hill marking the final approach to the school. We slowly pulled into the parking lot and the automatic doors swung open, releasing the crowd of students and signaling the start of a new phase in my life.

As sixth graders, we were assigned a homeroom as well as a locker with a combination lock in which to stash all of our school stuff. Somehow, I had the idea that every class would take place in a different room, but to my relief, I found that was not the case.

At least 30 students spilled into my homeroom, filling the seats. As the girl's soccer and basketball coach, our homeroom teacher, Mrs. Nelson, always dressed in a terrycloth warm-up suit with a gruff demeanor to match. Most days she sported a whistle around her neck. Treating the classroom like basketball practice, Ms. Nelson called the class to order, barking out names as she read through roll call and laid out the game plan for the year.

Although anchored by our basic homeroom, we rotated classrooms for science, P.E. and math. How I managed to make it through middle school math with a "B" average, I'll never know. In fifth grade, we went at our own pace, with each person moving along once he had mastered the material. This procedure turned out to be an appropriate segue into middle school math.

All the sixth-grade students were separated into classes according to their capabilities. Students like me who needed the most review and practice were relegated to the beginning math class. Everyone else was pushed up into the higher-level math classes preparing for pre-algebra. I sat in math class with my book in front of me, staring at the pages of fractions, multiplication, and division problems. I worked hard at math and tried to remember things that Carol had taught me in the summer class and tried to apply them. I kept several packages of graph paper handy and filled up my notebook so that I'd be ready for anything the teacher might throw at us. I managed to get through math that first year with a very basic understanding of the material and moved along to the next level.

English took on a whole different personality in middle school. Gone were the familiar reading and comprehension exercises that made school a safe and interesting place for me. Even if I failed miserably on a math question or got back a writing assignment dripping with red ink, I knew that if we could just read and talk, all would be right with the world again. Instead of the familiar and frequent reading assignments, we were assigned book reports, grammar, and punctuation lessons. Reading assignments were rare, simple, and relatively short. We were also required to give oral book reports in front of the entire class. Everyone stuck to the basics, summarized the plot, shared their own thoughts about the book, and stepped down as quickly as possible.

> **" I sat in math class with my book in front of me, staring at the pages of fractions, multiplication, and division problems. "**

The Source for Visual-Spatial Disorders

Britt: Things I Learned in Middle School, continued

Familiar with some grammar lessons from fifth grade, I thought that we'd just simply go over the components of a sentence and move on. I was eager for interesting writing assignments. As we progressed through the grammar rules, I came to realize that we were going to be spending a lot of time strictly on grammar lessons. Writing in middle school was not going to be as exciting as I had hoped.

In middle school, science became one of my favorite classes. Mr. Banks, the science teacher, had a wonderful reputation for being both tough and fair at the same time. I knew that I'd get along just fine with this teacher when he wrote his name on the chalkboard the first day of class using an odd mixture of both lowercase and uppercase letters. He went on to explain that he wanted us to ignore his writing style because the information was more important.

"Don't get caught up in how I write; rather, pay attention to what I write," he informed us. Mr. Banks' classroom looked like I had always imagined a science laboratory would look. Microscopes, astronomy charts, and periodic tables lined the room. Mr. Banks used demonstrations and activities to teach us different classifications of foliage. He handed out a list of fifty native trees and their descriptions. The assignment was to go outside, find each tree on the list, and pick a leaf, preferably one that had already fallen to the ground, as proof of our discovery. For several weeks, everyone in the class

> " . . . once I joined the band, I quickly realized that music was never going to be my strong suit. "

traded leaves and discussed the best places to look for the most difficult to find species. After collecting the leaves, we were tested, having to name the different types of trees by looking at and touching the leaf. Each leaf had a different shape and feel; some were fuzzy, some smooth, some rough or prickly. I ended up receiving an "A" on the test.

As we moved on to study astronomy, gravity, orbits, and planets, Mr. Banks used a method perfectly suited to my learning needs. Outside on the athletic field, we set up a solar system built to scale using inches instead of miles. The sun sat firmly in the middle of the field and the rest of the solar system stretched out past the front parking lot almost to the main street. We were amazed to see how the planets related to each other and to observe the huge distances in between them.

Middle school offered a choice of music electives, either band or choir. Although I enjoyed singing, I liked the kids in the band more than I liked the students in choir. Choir had the reputation for being the easy way out, which only further convinced me to take up the clarinet. I told myself that I could learn to read music with the right teacher, but once I joined the band, I quickly realized that music was never going to be my strong suit. As I practiced every night after school in my room, my brother would let out a tortured moan and ask me to stop.

One night, as I opened the clarinet case and started to put the instrument pieces together, he complained, "You sound like a sick goose." No matter how hard I practiced, or how much I practiced, I still had difficulty reading the music and keeping tempo. The students in band were seated according to instrument, and every few weeks, we had competitions to determine the best player in each section. Just as in an orchestra, the best player is referred to as *first* and so on until all of the players were placed according to their ability. Every time we had a competition, the other players rooted for me and hoped that I'd get a chance to move up. Three of us played clarinet, and during my entire time in band,

Britt: Things I Learned in Middle School, continued

I played third clarinet. I considered quitting, but I enjoyed the kids in band and wanted a chance to try to learn music. Besides, dropping band meant that I would have to join the choir. After my problems with piano, I was also a little afraid that either Dad or Mom would get mad at me for quitting.

I don't remember our band being very good. In fact, Mom has reminded me that our concerts, intended to show our progress to parents, were pretty awful. I'm sure that there were some students who really did play well, but the rest of us drowned them out.

In seventh grade, math proved to be a challenge. Fractions and percentages still thwarted me, especially multiplication and division of fractions. Having to turn them around in the process of solving the problem constantly confused me. By the time I had finished the problem, I had reversed the top portion of the problem with the bottom and ended up with an absurd answer. Each time I went up to the teacher and asked a question, the teacher just went over the instructions in the book again as if more repetition would help.

By eighth grade, I was in pre-algebra. This instructor called students to the chalkboard to solve problems. We were given problem sets every night and were expected to share our answers the next day as we reviewed the homework. Positive and negative numbers were a sore spot, and I flinched any time we had to work with them. We worked with basic algebraic concepts such as variables. I liked the idea of not knowing what the letter stood for since understanding numbers was difficult in the first place. Plus, I was thrilled to see letters (something I could relate to) instead of a foreign symbol or bizarre fraction. I pulled myself through pre-algebra and thought that if I could survive three years of math on steady ground, I'd be ready for algebra.

Both seventh- and eighth-grade English revealed much of the same dull course work as sixth grade with various interludes, such as the visit to the school library for the standard lecture on the card catalog and proper library usage. The librarian passed around the same "pop-quiz" at the end of her talk every year. Having spent a lot of time in the University library waiting for Mom to finish a meeting or talk to students, I knew my way around a library and resented the librarian's ridiculous questions.

> **"** *Grammar and vocabulary were stressed in this class and counted for exactly half of the grade.* **"**

My eighth-grade English class was held in one of the portable classrooms located on the side of the school. The small classroom windows let in miniscule amounts of light, bathing the entire room in a dusty yellow glow. Grammar and vocabulary were stressed in this class and counted for exactly half of the grade. The teacher had us memorize both the meaning and the spelling of the vocabulary words. During the vocabulary portion of the tests, the teacher called out the words and we were expected to write them on sheets of paper. Then we moved on to the grammar section of the test. After getting our tests back, I would cringe at the red marks and the "C-" scrawled on the paper.

For the second half of the class, another teacher in the classroom directly next door taught us writing. Thankfully, this portion of the class balanced out my overall grade. For one assignment, we had to take a current newspaper article and re-write the events as they might have been reported 100 years ago. All of these creative writing assignments and essays proved to be fun and easy, and I was successful with them.

ஐ ஐ ஐ

Carol: Middle School Academics

Early on this crisp November day, Julie called to say she wanted to come in to talk. As she slipped into my office chair, I asked, "Has something happened with Britt?" I had not seen her for nearly a year.

"No, Britt is fine," Julie replied.

Now in the eighth grade at Harbor Middle School, she had "bumped and slipped through the past two years," in her mother's words.

"But," she explained; "now John is in the sixth grade at the same school, and his courses are as challenging and exciting as Britt's are not." Julie's worries tumbled out. "Britt and John always seemed so similar in their interests and ability except for Britt's problems with math. Why was John chosen for the gifted program while Britt was not?"

Julie understood the testing process. Her questions were much deeper. Was Britt really as capable as she seemed to her mother? If the school could challenge John, why not Britt? Why was English so terribly dull? Britt loved to read and to write, but she hated English class, probably because at least half the class was devoted to grammar. Why did Britt have these endless grammar exercises when research on composition had demonstrated that these exercises would not improve a student's ability to write standard English?

She paused, looked at me, and fell silent. I cleared my throat. "Julie, I cannot really say why they set up their curriculum this way, but you would find the same arrangement in most schools—public and private. I agree with you that research indicates that these drills have poor transfer, but applications always lag behind research."

Julie looked doubtful. "But why then?" she said.

"Educators have changed their minds frequently about what to teach 12-, 13-, and 14-year-old

> **" *Why was John chosen for the gifted program while Britt was not?* "**

students. They have changed the length of class periods, the format of the day, and the availability of extra-curricular activities."

She nodded, the concerned look still on her face. I went on, "Everyone agrees that children in this age group have arrived at a very important stage. Public school districts and private schools have responded to this stage in a variety of ways. Most try to build in lots of activity and a wide variety of classes."

I added, "High school is not going to be any different. Most of them offer core courses that range in difficulty from very easy to very advanced. But the entry process—well, the fact is—placement tests will discriminate against Britt. They do not provide for the student who is very intelligent, but who has pockets of disability."

I went on, "Britt has some problems fitting into any standard system. She is not a typical student. Both her abilities and disabilities are a challenge for schools.

"Your challenge with a public school is to adapt a system, a big system, to the needs of a very intelligent, learning-disabled youngster."

"Well," said Julie as she snapped her notebook shut and stood up, "I'm going to talk to some administrators at Britt's school."

I thought, "Go for it, Julie, but you are not going to change that system."

༄ ༄ ༄

Julie: Middle School

Soon after my conversation with Carol, I happened to read an article about public school support for children with disabilities. Britt would surely qualify and I certainly needed funding support for her. Maybe this support would be the help that Britt needed. I imagined the school would provide educators like Carol who could provide Britt with help and at the same time, extra challenge. I asked Carol and Britt's pediatrician to send files to her school, and I made an appointment with the vice-principal.

Wearing my "school-appointment-black-suit," I waited in the reception area for Mr. Anderson to see me. I watched the kids in the hall shuffle past and saw, perhaps for the first time, that most of the boys were twice as tall as Britt. The bell rang and a tall, slender man appeared in the door. "You must be Mrs. Neff. Please come in," he said motioning to a chair across from his desk. "How can I help you?"

"Have you had a chance to review Britt's file?" I asked.

"Oh, yes," he said, shuffling some papers on his desk. "You probably saw that she has been diagnosed with Turner's Syndrome and a serious spatial learning disability."

"Yes, I did see that, and I talked with four of her teachers. They all said she was doing just fine in her classes."

"But she needs some extra help to do 'just fine'; I'm wondering if she qualifies for federal funding under one of the programs. She does have a disability, and I wanted to find out what kind of help existed."

"Well, yes," he said, looking at the papers again, "she definitely qualifies, but all we have here is Title I which means she would be put in a remedial class with remedial students."

"But she doesn't really need remedial help. She needs help learning the material that is part of the curriculum in regular classes," I tried to explain. "Just repeating the same material won't help. She's very bright and capable of doing more in some areas, but she needs some support in others."

"But those classes are the only ones we have. I wish we had something else, and frankly, I would not recommend putting Britt into those classes. Britt has a good attitude and sees herself as a successful learner, and as you mentioned, she does very well in some areas. I'm afraid that being surrounded by students who don't have those qualities would be a negative experience for Britt."

"So there's nothing the school can do for Britt even though she's entitled to it?" I knew the answer before the words left my mouth. I wanted to ask about the gifted program for Britt, but from this conversation I could tell that the route to the gifted program was through testing and Mr. Johnson was not flexible enough to see an alternative.

"I'm sorry I can't help you," he said as he stood up to signal the end of the conference.

As I walked to the car, hot tears filled my eyes. The system was so unfair. Britt was denied assistance because she wasn't a troublemaker, because she had a good attitude, and because she worked so hard. Help existed in the community, but why didn't it exist in the schools? I was angry and disappointed.

As I drove, I calmed down and analyzed the appointment. The good news was that Britt's teachers saw her as a successful student even though she was

> **"** Britt was denied assistance because she wasn't a troublemaker, because she had a good attitude, and because she worked so hard. **"**

Julie: Middle School, continued

having trouble in her math and English classes. English class had become a problem because of the grammar component. The middle school English curriculums taught parts of speech, sentence diagramming, and spelling in addition to reading and composition. Recognizing parts of speech and diagramming are both spatial tasks that have nothing to do with a student's ability to write Standard English. This was apparent in Britt's compositions. She could write clearly and grammatically correct compositions even though she could not identify parts of speech. She continued to have trouble with punctuation and spelling—both of which are also spatial tasks.

At home, I could help her make the corrections; but in class, she lost points for those errors. On the grammar and spelling tests, she received "C-'s" and "D's." By this time, Britt better understood her own disability and could understand when I told her that these tasks would always be difficult for her. Rather than worrying about them, she should be proud of her ability to read, write, and speak. Still she hated the "C+'s" she received in English, which reflected the averaging of the "A's" and "B's" in reading and the "C's" and "D's" in grammar.

Through middle school, Britt excelled in the things she was good at. She enjoyed social studies because it allowed her to bring together different kinds of knowledge that she had gained from her extensive reading. However, the books and class discussion were below her ability, and she longed for more challenge. To some extent, special projects helped (Her project on the country of Norway was the Grand Prize at the school exhibition), but she longed for more challenge in her day-to-day work.

Britt also did well in science, which was hands-on and tied interesting experiments to everyday life. But her grades seldom reflected her ability or her need for more challenging work. Why couldn't schools provide help and more challenge at the same time?

This problem became even more apparent because Britt's brother John started sixth grade at the same school and tested into the Gifted Program. From the first week, John was reading thick books full of new and challenging ideas. One afternoon, Britt was thumbing through one of John's books, looking increasingly dismayed.

"You know," she said, "I can read this book, and I could have read it in sixth grade. I'm a better reader than John is, but they won't let *me* into the program because I can't do math very well. It's so unfair." Before I could answer, Britt had stomped out of the kitchen. I could only hope that in high school, Britt would find the right balance of intellectually challenging courses and help with basic math concepts.

During the early spring of Britt's eighth-grade year, two things happened that would change our lives. First, I fell in love and decided to marry a wonderful man who I had known for several years. My new husband John accepted both Britt and my son John. He brought one child to the marriage, a daughter Sarah, who was six years younger than Britt.

Secondly, eighth graders and their parents were invited to visit the local high school to learn about the programs.

On visitation night, Britt and I attended the first session in an English classroom. Dressed in a red blazer and coordinating skirt for the event, the teacher was personable and enthusiastic. She talked about the reading and writing curriculum, and announced that the students would continue to

> **" *The books and class discussion were below her ability, and she longed for more challenge.* "**

Julie: Middle School, continued

refine the grammar they had learned in middle school. The reading had been chosen to let every student succeed.

"We don't want to turn kids off when they first get to high school, so we start with some of the favorites," she said. "We'll read several advanced Judy Blume books, some adventure stories, and *The Diary of Anne Frank*."

Britt leaned over and whispered, "I read all the Judy Blume books in sixth grade."

"I know," I whispered back.

I looked up at the spelling and vocabulary words on the chalkboard. *Occasionally* was spelled *ocasionally*, but Britt didn't notice and I didn't care.

We went to the French class next where Madame Langlow said she was happy to meet us, but foreign languages were for the "advanced" ninth graders.

"It's just too overwhelming for most freshmen," she said.

"But how do you determine who's ready?" I asked.

"Those in the gifted program almost always qualify, and then there's a standardized test that we can give if other indicators of success in foreign language are good," she replied.

"So what are the indicators?" I pressed on.

"Well, usually grades in English and the standardized achievement tests that all eighth graders take. These are excellent predictors," she said smiling.

Britt whispered again, "I guess that leaves me out."

Walking past the student activity tables, we headed back to the car in the drizzle, deciding to skip the cookies and fruit punch in the cafeteria.

Britt was quiet.

"Well what do you think of high school?" I asked with a cheery voice.

"I think high school stinks," Britt replied, tears welling in her eyes. "It's just more of the same. I hated it and I don't want to go to school there."

Although I agreed with Britt, I didn't know of any alternatives. We drove home in silence. Britt's response to the high school and my own misgivings about the curriculum led me to ponder what other choices there might be. What help could there be for the unusual learner?

When I got home, young John was already asleep in bed and I found my husband John in bed reading a big stack of magazines.

I explained what had happened at the school and how disappointed Britt was. John encouraged me to pursue other options for high school.

"We don't even know what's out there," John said. "We need to look around before we make a decision." I went into the bathroom to wash my face, feeling the high school problem was more complex than John could see, but also realizing we would need to explore the options.

> " Britt's response to the high school and my own misgivings about the curriculum led me to ponder what other choices there might be. "

ಆ ಆ ಆ

The Source for Visual-Spatial Disorders

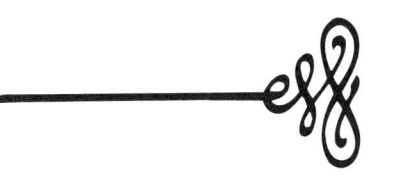

Action

"The best way out is always through."

Robert Frost
1874-1963

Julie: Transition to High School

In the days after the high school information night, I thought a lot about Britt's negative reaction to the local high school. As John suggested, I started asking friends about other choices. Two private high schools emerged as possibilities: one a co-ed Catholic school, and the other, Annie Wright, a girls' school with Episcopal affiliations.

The Catholic school had a good reputation and was less expensive, but I felt I needed to look into both. I wrote for information and made arrangements for Britt to visit the schools.

Both schools felt better to us than did the local public school because they were less interested in pre-judging students based on standardized test scores and both had a curriculum that would allow for Britt to take advantage of some of her strengths.

While the Catholic school provided more social opportunities, the girls' school had smaller classes and a close-knit community of girls and teachers. After the campus visits, it was clear that the girls' school was Britt's first choice. She liked everything about the school, including the fact that her aunt Jill taught there. She even liked the idea of wearing a uniform. Although I still had my doubts about the girls' school and the cost of tuition, John was convinced that the small classes and challenging curriculum would be worth the financial sacrifices we would have to make.

Over the next weeks, we filled out the application forms for both schools and gathered letters of recommendation. Britt put her application essay into the packets and dropped them in the mail. "Well, I did my best," Britt said. "Now we just have to wait."

Within a week, the Catholic school called to say that they would have a place for Britt in the freshman class. Two weeks later, the acceptance from Annie Wright arrived. We had already decided that if she was accepted at Annie Wright, she should go there. We were relieved and delighted with the acceptance.

Once the excitement of the acceptance had passed, I began to worry about how Britt would meet the challenges of this new curriculum. I believed that Britt could excel if she were given interesting and challenging material and if the emphasis was taken off skills and drills, but I had to admit I didn't know for sure. I worried that I might be setting her up for failure. I explained my worries to John who asked me what I wanted to do about it. I had to admit I didn't know. The next day I called Carol to see what advice she might have to offer.

Carol suggested that Britt spend a few hours at the clinic toward the end of the summer, brushing up on some of the skills she had learned in middle school. This seemed like a good idea to me, but I wanted to give Britt the chance to decide for herself.

> **"** Once the excitement of the acceptance had passed, I began to worry about how Britt would meet the challenges of this new curriculum. **"**

Later that morning, I picked Britt up from swimming lessons and explained my talk with Carol.

"That's fine," Britt said. "Late in the summer sounds good; I won't forget what we do before school starts up again."

I couldn't help but think back to the first time I had suggested a summer class at the clinic and to how angry Britt had been. Britt interrupted my thoughts, "When am I going to order my uniforms? How many do you think I'll need? And I also need sturdy navy blue shoes, extra white blouses, and socks, but

The Source for Visual-Spatial Disorders

they have to be navy blue or red or white . . ."

"Okay, Britt, I know you're excited about school, but we have plenty of time to get the uniforms, and the socks, and the blouses," I said. "More than a month. When we get home this afternoon, let's make a list of what you'll need and what you have. Then we'll plan some shopping days, including a visit to the uniform store."

In early September, when the first day of school arrived, I drove Britt to school, stopping in front of the long, three-story brick building. Britt was so excited! She jumped out of the car, gave me a quick wave, and hurried down the long walk to the school. What would this place hold for Britt? I hoped for the best.

Britt's excitement continued through the next two weeks. She liked the girls, she liked the teachers, she liked the uniforms, she liked the food in the cafeteria, she liked everything about the place. As a result, so did I. It was wonderful to see Britt so happy.

As time passed, the homework increased, and it became clear that the transition to the new school would not be as easy as it might have first appeared. Increasingly, she went directly to her room to study when she got home from school. After dinner, it was more studying until bedtime. The school days were longer now too—from 8 to 3 with six or seven subjects depending on the day of the week. Sometimes Britt stayed late so she could use the school library and get a jump on a project. She often fell asleep in the car on the way home.

But Britt never complained about the homework. She enjoyed making the connections between literature and history that grew out of her Civilization course, and she was pleased that she was learning French, often trying out what she learned on me.

If she asked, I helped with the editing of papers and served as an audience when she needed to practice something orally. For the most part, Britt took charge of her homework, working hard and seldom complaining. I think we both wanted to believe that her learning problems were behind her.

Unfortunately, as the first papers and tests were written and then returned, it became clear that Britt's struggle would continue. Her teachers had high standards for content and style, and Britt would have to find ways to cope and adapt if she wanted to stay at Annie Wright.

ಌ ಌ ಌ

> **" For the most part, Britt took charge of her homework, working hard and seldom complaining. "**

Britt: Getting Ready

Anticipating a rigorous academic schedule and difficulties with algebra, Mom suggested I take a course over the summer between eighth and ninth grade to prepare me for some of the challenges I'd face in high school. I don't remember looking forward to this class, but I knew I needed to brush up. Since I had worked with Carol, I knew what to expect. Carol was not available to teach that summer so Margaret, an instructor at the clinic, took her place. My excitement about going to a new high school overshadowed any reservations about this summer math class.

I knew that I'd be taking algebra in the fall, so over the summer we worked on equations, fractions, decimals, percentages, negative and positive numbers, and graphs. We made keys to remember the algebraic equations and worked on strategies to remember how fractions and decimals are treated in the equations.

Eighth-grade English had been difficult, and I still struggled with spelling and grammar. My aunt Jill, the art teacher at Annie Wright, told me that there would be a lot of writing in all of my classes, and I realized that I needed to work on mechanics in order to do well.

Margaret suggested that we work on punctuation, spelling, and grammar through journaling and handed me a blank book for keeping track of my thoughts. The idea behind keeping a journal was to write a little every day and bring the writing to class with me. In class, we reviewed aloud the passages I had written, looking for misspellings and words I had left out. I loved the idea of having my very own blank book just waiting for my words and thoughts to fill the pages. It was mine, and I could write anything I wanted to write.

Feeling empowered at the thought of having my very own journal, I gladly took on the challenge and wrote about everything from people I knew to a description of our house. I didn't feel like I was being judged, and I enjoyed the assignments. The journal was successful because it helped me evaluate my own writing and find out what was missing.

Although the summer class was never as much fun as the Constructions class, I was motivated by knowing that school would be hard in the fall. Although I was thankful that the school had chosen me, part of me thought that I should not have been accepted, and I was scared of failing!

The week between finishing the class and starting school went by quickly. I spent part of that week helping my aunt arrange her art history classroom. Dozens and dozens of old photocopied pictures mounted on cardboard and dusty slides left behind by her predecessor lined every inch of the room. The classroom faced east and by late morning the August sun had made the room unbearably warm and stuffy.

Seeking refuge from the hot classroom, my aunt gave me a tour of the school. Since I had attended Annie Wright in kindergarten and first grade, I recognized my old classrooms as we wound around countless corridors and made our way up several flights of stairs.

As we walked through the hallways, my aunt gave me some history and background about the school.

"Annie Wright was founded in 1884 by Charles Wright, a railroad tycoon from Philadelphia, and named after his daughter." My aunt pointed to the life-sized portrait of a woman that hung in the main stairway.

> *My excitement about going to a new high school overshadowed any reservations about the summer math class.*

The Source for Visual-Spatial Disorders

"The original building was burned to the ground in 1900 and rebuilt around 1920," my aunt said as I stopped to inspect a few old pictures hanging in the hallway.

Next we passed a room shimmering with colored light from the stained glass lining the walls. "This is the chapel. Everyone is required to attend three days a week. Father Berge leads all of the services," Aunt Jill continued. "We also have a 'Lessons and Carols' service every year before Christmas. I think that's my favorite service because the chapel is always decorated with flowers and candles."

I wondered how I could ever find my way through such a massive and confusing place. But as we talked our way through the different areas of the school, I became comfortable navigating the labyrinth of winding hallways and stairwells by relating each part of the school to my aunt's stories.

After seeing the main hall, dining room, great hall, upper school and lower school classrooms, chapel, library, gym, pottery studio, art studio, music room, pool, and upper school lockers, we concluded the tour in the main hall. I had a wonderful sense of belonging and a reinforced sense of excitement about all of the classes I'd be taking.

On the first day of school, I got up early and put on my new

> **"** *I had a wonderful sense of belonging and a reinforced sense of excitement about all the classes I'd be taking.* **"**

uniform for the first time—a crisp white cotton shirt, knee-length plaid skirt, white knee socks, and navy blue loafers topped off by a red necktie and navy blue sweater. I was ready.

☙ ☙ ☙

Britt: Annie Wright

As I look back over my report cards from my freshman year in high school, I am simply amazed. They worked us hard: Art History, French I, Studio Art, P.E., Algebra, English, Social Studies. The year before in my middle school, I had taken Math, English, Social Studies, Band, and P.E. Not only did I now have seven courses instead of five, but they were also all taught at a much higher level. Nothing could have prepared me for the sheer volume of work I encountered at Annie Wright. Moving from middle school to high school marked an immense transition in terms of the subject matter of writing assignments and the amount of writing required for classes. Intensive writing pervaded subjects I had never thought of as writing courses—Music History, Art History, History, and even Biology. Caught off-guard by the number of writing assignments, I worried whether I'd be able to handle Annie Wright and its new demands.

We settled into our weekly routine. I was happy sporting my uniform and felt important with all of my new books the lady at the bookstore had handed me with a smile, asking me to sign a form stating that I had purchased my books. There were so many books that I could barely fit them into my book bag, and I struggled to get them down to my locker. No protective covers were required, and I marveled at the fact that I could keep the books even after we finished the school year.

Clutching my schedule in one hand and books in another, I headed for Madame Maxwell's French class. Learning from my aunt that Madame Maxwell was a visiting teacher from a small town in France, I imagined the worst. Visions of a strict disciplinarian with glasses perched precariously on the end of her nose, hair pulled tightly into a bun at the back of her head, holding a ruler in one hand, and never speaking a word of English invaded my dreams. Instead I found a friendly, petite woman with stylish brown hair casually standing at the chalkboard conversing with the students as they settled into their desks.

She must be Madame Maxwell, I thought, heaving a sigh of relief.

"Bonjour!" Madame Maxwell greeted the class.

"Welcome to French class." Madame Maxwell switched to English in order to introduce herself and explain the text book we'd be using and the listening lab.

"Any questions?" Madame asked and looked around at the students.

"Will we be allowed to use English and French, or strictly French?" I asked, wanting to ease my only reservation about this class.

"Ah, bon! That is a good question. Until you build a vocabulary base, we will speak both languages." And with that clarification, Madame Maxwell instructed us to open our books to the first lesson.

I enjoyed learning French. We learned simple phrases and vocabulary for the first year. I learned to speak with an excellent accent, but spelling became even more challenging than English. I learned that vocabulary tests asked for three pieces of knowledge: the meaning of the words, their spelling, and whether the word was feminine or masculine. I managed the spelling by connecting my own images to the vocabulary (e.g., "le chat" = "c + hat." See a cat with a hat.) I also looked for smaller words within the vocabulary words to remember the correct spellings. And, when I put these two strategies together and practiced aloud, I sometimes even amazed myself with what I knew.

Later that year, as I progressed in French, I encountered verbs and conjugations. Even though

I knew the meaning of the verbs and the pronunciation, I struggled to learn the uses of regular and irregular verbs.

At this point, studying French became the equivalent of being on a disorienting ride at the county fair. The rows of verbs varying only slightly made my mind spin. First of all, I had to learn at least twelve different conjugations for every verb. Secondly, I had to keep the forms of the verb consistent with the verb tense and match the person with the proper verb form. Most of the verbs ended up melding together along with the grammar rules. However, in conversation, I could repeat the correct pronunciation and recall a great deal of vocabulary. Despite fighting to keep up with the written work, I began to speak French.

Instead of having separate classes for English and History, the freshmen had a class called Civilization, which combined the two disciplines into one. Ms. Blake, the Head of Upper School, taught the literature portion of the class two days a week, and Ms. Soucey taught the history portion the other two days. One day a week they taught together.

On the first day of class, Ms. Soucey and Ms. Blake worked together introducing the students to both sections. A group of 25 students crammed into the classroom and waited silently for the bell to ring. Ms. Soucey stood at the front of the classroom putting handouts into neat piles across the length of the desk. She paused to adjust her wire glasses and frowned as she counted the students seated around the room. Ms. Blake stood facing the chalkboard, absorbed in her writing, oblivious to the anxious eyes focused on the front of the room.

As the clanging of the bell subsided, Ms. Blake woke from her trance and turned around to face the students. She was the perfect image of a headmistress of an all-girls' boarding school. She had long, gray hair tied neatly into a bun at the back of her head and wore a long, flowing skirt and crisp blouse which gave her an air of benevolent authority. She picked up the roll sheet and began to work her way down the list of students, carefully enunciating each name. As each student answered, Ms. Blake peered over the reading glasses set squarely on the end of her nose and gave a quick nod.

"Ms. Soucey and I are going to hand out the syllabus for the entire year, and we will go over it together so that there will be no misunderstandings about what we expect," Ms. Blake said, taking one of the stacks of paper from the desk. As I received my syllabus, I saw the lists of books and stories that we were going to read, and my heart leapt at the thought of an entire class devoted to these stories.

"As you can see, this class will be intensive. We are going to cover a lot of ground in a very short period of time. I know that some of you have an advantage because you already know some of the material. However, most of the reading

> **"** *I had a new start and teachers who expected me to do as well or better than anyone else in the class.* **"**

will be new for everyone, and I expect that all of you will be at the same level before long," Ms. Blake instructed.

At that moment, I knew that I belonged at Annie Wright. I had a new start and teachers who expected me to do as well or better than anyone else in the class.

We continued with our discussion of the syllabus and discovered that we would be covering everything from the *Epic of Gilgamesh* to Shakespeare's *A Midsummer Night's Dream*. They told us that at the end of the year, there would be a large paper that would require us to define *civilization*. Thinking about the paper made me

Britt: Annie Wright, continued

nervous. How will I ever finish such a daunting project? Never having written a paper that required me to go outside of an encyclopedia, I wondered where I would begin.

I proudly brought the bright blue 1,000-page *Adventures in World Literature* Ms. Blake had assigned to class for our first official meeting. I held the book with both arms, and my heart sang as I realized that this book was crammed with stories and plays. This was what I'd been waiting for—an entire class devoted to reading and discussion. Ms. Blake told us to flip ahead to the section on Greek Literature, even though ancient Sumeria occupied the first few chapters of the book. We started with Homer and sections from *The Iliad* and embarked on a lengthy discussion of Greek mythology and history.

I wanted to participate in the class discussion, but, at first, I kept quiet so that I wouldn't appear dumb. Haunted by insecurities from elementary school and middle school, I wondered what I could possibly contribute to the conversation. Except for three or four of us, the other girls in my section had all been at Annie Wright since they started school—"lifers," as they were called. They obviously had a better grasp of the material and knew what to do when it came to writing a paper or extolling the virtues of the text.

"What do you think Homer is trying to convey in the passage?" Ms. Blake looked directly at me and waited for me to answer. Inexperienced at writing or talking about the merits of world literature, I tried my best to answer questions posed by the teacher.

"Well, umm," I paused and looked at my book to make sure I had the right page and paragraph. "Maybe, he's talking about courage and loyalty?" I stammered, turning bright red.

"Yes, I think that's a good answer. But how does that relate to the characters directly? Anyone?" Ms. Blake asked.

The bell reminded us that it was time to go; Ms. Blake sent us home with several chapters to read for the next day, telling us that tomorrow's discussion would be on translation, and we would look at two versions of the same text. Relieved that I survived with only a minimal amount of embarrassment, I started to feel better about the new demands that were being placed on me.

At our next meeting, we talked about point-of-view and how the author's translation of the text can be influenced by personal and cultural factors. Our assignment was to paraphrase a portion of the text and rewrite it using familiar language. Secretly, I was scared, and I knew that I was in over my head. How could I handle all of the writing assignments, especially if I didn't understand what the teacher wanted from me?

Unable to think of any other way to do the homework, I slowly and painstakingly picked through the assignment word by word. First, I underlined all of the words I didn't know. Then I looked up each one in the dictionary and wrote down the definitions. Then I outlined the different actions in each paragraph to help me keep the plot intact. Finally, I looked at each sentence, keeping the main ideas in mind and searching for alternative words in the thesaurus.

Once I finished my draft and I was satisfied that it all made sense, I copied everything onto a clean sheet of paper with an erasable pen. Three and a half hours later, I printed my name at the top of the completed assignment, too tired to worry

> *" Haunted by insecurities from elementary school and middle school, I wondered what I could possibly contribute to the conversation. "*

Britt: Annie Wright, continued

about what type of grade I'd receive on my first high school paper.

Monday morning, Ms. Blake walked into class and said, "I have your papers to hand back, but before I give them back to you, I want to read from a paper that I think is particularly good."

Expecting to hear Ms. Blake read a paper from one of the lifers, I did a double take when I realized she was reading my words. I held my breath, still not believing my paper was being read. That one moment gave me the confidence to keep going and tackle even harder projects.

For the next assignment, we were expected to write an essay. My eyes grew large with dread when I realized that, unlike the paraphrase, I had no model. I was lost. Without a model of a proper essay, thesis, or argument, I worked on the paper one sentence at a time, struggling to make my words coherent. My inability to distinguish the logical pattern of an essay made creating a structured paper impossible, and I turned in an assignment that was no more than a collection of random thoughts.

Apparently I was not the only student who did not understand the concept of writing a well-organized essay. After we completed our first attempt, Ms. Blake decided we needed to learn how to write a proper five-paragraph essay. After she explained using concrete examples and we asked questions, I realized for the first time that there was a way to give order to random concepts! It finally made sense to me.

What I learned about essays in English applied to History even though history papers were somewhat different. Most history papers were longer and more involved than English papers. We were expected to probe primary and secondary sources, organize the material from each, and develop an original thesis. We talked about the papers in class, which gave me an idea of what the teacher expected and other ways of thinking about the materials.

Ms. Soucey's reputation for being the hardest teacher in the school was legendary. At lunch one afternoon I sat with a group of students discussing their assignments. I listened as the sophomores and juniors compared the papers they were writing for Ms. Soucey.

"Ms. Soucey is the hardest grader in the school. Last year I wrote a paper for American History. She even took off points for a simple typo in the bibliography," the junior told the table.

"Yeah, last year she failed one of the girls in our Civilization class just because she turned in a paper ten minutes late!" the sophomore added.

I felt a knot form in my throat as I tried to stay calm and not think about the approaching paper. I lost my appetite and left the lunch table feeling more nervous then ever.

As the end of the year approached, the deadline for the final Civilization paper neared. I had no idea how I was going to include all of the materials and ideas we had been discussing into a ten-page paper. We had many sources and a very broad topic: "Define Civilization." The sheer mass of information we had learned scared me, not to mention that I had never written a paper this long before or developed such a complicated and sophisticated thesis. In addition, I had to tackle editing and make sure that my sentences made sense as I got them onto paper. I kept thinking about the discussion I had heard at the lunch table.

> " *We talked about the papers in class, which gave me an idea of what the teacher expected and other ways of thinking about the materials.* "

Britt: Annie Wright, continued

I also had to learn how to divide my time so that I did not have the entire paper to write the night before the due date. I had trouble thinking about how much time each step in the writing process would take. I had never done research, made an outline from notes, or composed a bibliography or endnotes. Each part of the paper-writing process took longer than I expected, adding to my frustration and panic about the ensuing deadline. I also had to make sure I finished homework assignments on time and studied for final exams in all of my other classes. I arranged to have my paper typed and just barely managed to get it back from the typist before I turned it in to Ms. Soucey. Embarrassed, I handed in the paper and hoped that it was good enough.

Ms. Blake and Ms. Soucey did not receive my paper with open arms. Editing mistakes were everywhere, making the text hard to read. My ideas were lost in the mire of mistakes. Clearly I had struggled with this abstract, diffuse, and vague topic and had tried to put my ideas into the paper. Ms. Blake and Ms. Soucey gave me a passing grade.

> **" Each part of the paper-writing process took longer than I expected, adding to my frustration and panic about the ensuing deadline. "**

That fall I also met high school algebra. Luckily, the summer class at the clinic had helped me recognize the algebra equations. I chose a seat in the front of the room as I usually did. The first week we had our initial in-class assignment, and I struggled to finish in the allotted time, but I couldn't do all of the problems.

It wasn't that I didn't understand the algebra; it was that I couldn't do it fast enough. The comments on my report card revealed my struggles. "Britt has worked very hard to maintain a 'B' average in her class work. She does need to push herself to work faster, as she often does not finish in-class assignments or tests." Of course, a "B" was a great grade for me; in fact, the best I could have hoped for, but I felt she saw me as a student who was lazy, not working hard enough or fast enough. I completed the first semester of the course with a "B" average, but it had been a constant struggle.

Second semester began and my woes with math were not over. Ms. Brown grew increasingly impatient, and we began to cover unfamiliar material. Quizzes consisted of nightmarish algebraic equations, and most of them included fractions or decimals. Tears came to my eyes as most tests came back with a "C-" or "D." I had to do better if I wanted to move on to geometry next year, and I knew I had to take geometry to graduate and qualify for admission to college.

ಐ ಐ ಐ

Carol: Cracking Algebra's Code

Britt's cheery voice on the phone announced that she needed some help. She was near the end of the spring term of ninth grade in what had proved to be a challenging year, and was calling to reserve a place in my summer schedule.

I heard her sigh as she said, "It's algebra. I'm passing Pre-Algebra, but not by much. I've been too busy with all my other subjects to do extra work. But I need to do more this summer—in fact, I want to."

Before I had a chance to reply, Britt laughed and added, "I can't believe I'm saying that, but it's true. I want to do math with you this summer."

After agreeing upon a time, Britt said a hasty goodbye. She had a paper to finish for History. Her schedule this year had been extraordinary. She had successfully made the transition to demanding academics and seemed to relish her classes. Well, most of her classes. Pre-Algebra was "a drag," as she described it. What could we do to get her ready for algebra? I started planning.

Britt had made progress in understanding the number system since the early days, pursuing the mysteries of subtraction. But math depends on spatial relationships, which are hard for Britt. No, *hard* was not the right word. *Mysterious* was a better word to describe the demands of spatial tasks.

I knew where to begin. We needed to build bases for her images as we did in the Constructions class. And we needed to talk through the concepts of math before she could substitute "x" for numerical values. Britt had shared how helpful developing a specific framework for essays had been in English class. Now we needed to make sure that she had an equally reliable framework for new concepts in math.

> " *We needed to build bases for her images as we did in the Constructions class.* "

Our cool offices were inviting on the hot summer day we began work. Britt arrived for her class in blue cotton shorts and a T-shirt. She looked comfortable and relaxed. "You look like a person who is ready to master decimals, fractions, and percents," I said as I put down a stack of papers and math books.

She said in return, "We are starting with decimals, fractions, and percents? Good. Let's put off algebra as long as possible."

"Sorry, Britt," I replied with a smile, "but that is algebra. What is more, they are all the same thing—just said in different ways."

Britt nodded, but looked doubtful. "If they are the same, why do we need all of them?" she asked.

"Good question, Britt," I replied. "They are used in different places. Like money . . ."

"Uses decimals. I know that," Britt finished my sentence and added, "And teachers use percent for grades. 'You only got 56 percent of the math problems correct,'" Britt answered, mimicking her Pre-Algebra teacher's voice.

We reviewed pivotal pieces of math to make sure she could put each of these processes into her own words. She remembered a lot from her summer class after fourth grade. That class, and all the work Britt and Julie had done in the kitchen, had clarified many basic concepts. Now we needed to translate these concepts into Britt's own descriptive language so she could transfer the ideas to algebra class. Fractions such as 2/3 were an especially confusing concept to describe. Britt needed to focus on the meaning of the *two* in relation

The Source for Visual-Spatial Disorders

Carol: Cracking Algebra's Code, continued

to the *three* and 2/3 in relation to the whole. She also needed to explain the relation of fractions to decimals.

We started with money, a concept she already knew even though she wasn't aware of it. Certainly she had never put money and fraction relationships into words. "Britt," I asked, "What is one-half of a dollar?" I wrote ½ on the paper and slid it in front of her.

"Well," Britt looked puzzled, "fifty cents of course." She wrote 50 below the ½.

"Remember, money is written in decimals," I said. Britt looked thoughtful, and then put a decimal in front of the 50.

"I'm not used to thinking of 50 cents that way, but I guess that would be the way to write it," she said.

"Exactly," I assured her and added, "Now divide the bottom number into the top number."

With just a few hints, she remembered that she had to put the decimal after the one and add two zeros.

$$2\overline{\smash{)}1.00}\quad .50 = \tfrac{1}{2} = 50\%$$

"Why, that's the same. You get 50 cents. Wow, people just don't go around saying that five dimes are 50% of a dollar, but they could," Britt said.

I smiled and gave Britt another fraction to divide. Now that she had framework and language for the relationship of fractions to decimals and percents, they lost their mystery. Soon she could describe how to convert fractions to decimals or percents in any order.

Over and over, Britt made discoveries about the meaning of numbers and the systems used to express numerical relationships. She quickly made the transition to more complex equivalents. A ream of copy paper on the table provided the next challenge.

"What if we took two sheets of paper from that package? What fraction and decimal part of the whole amount would that be?" I asked with a smile.

She read the label. "500 sheets in each package," she said. "That would be 2/500 for a fraction, wouldn't it? Wow!"

After correctly dividing 500 into 2, Britt exclaimed, "I don't think I've ever seen or heard anyone say the thousandth part of something except in a math book. I thought only engineers and my brother John understood that."

Britt's ability to describe number relationships in words was a huge asset. We used her language to develop a written reference sheet in her own words for fractions, decimals, and percentage. Each reference sheet was a single page divided into four sections. The sections had an example problem and the steps required to solve the problem. The four sections of the page were for addition, subtraction, multiplication, and division. Britt added a comment section at the bottom of the page.

We both knew her reference sheet for percentage was successful when we tested it using a math textbook. She discovered that her descriptive words would allow her to solve any percentage problem we found in the text. Fractions and decimal reference sheets passed the same challenge.

The summary sheets provided an overview of the whole picture for Britt (Mathematics: Operations, page 181). The overview, in turn, gave Britt control of the operation, and it gave her a point of entry for solving math problems. She had the tools to be in charge of her math skills.

Britt could also use the reference sheets to write her own story problems. Because story problems link math operations to real life events, Britt gradually realized the necessity of having an accurate image for the events before trying to do calculation.

Britt developed her four-point method:

1. Make the image (picture) for the story.
2. Put the picture down.
3. Put numbers in your picture.
4. Form questions from the numbered drawing.

These steps assured Britt that she understood events and could describe them before she complicated the process with numbers.

The number system bases the value of a number upon its position. For instance, 502 is a different amount than 520. Number position is a painful obstacle for anyone who does not perceive space accurately. Britt and I spent time talking through the use of commas to group numbers into threes. Certainly 1,111 was much easier for her to read and understand than 1111.

The concepts were usually obvious to Britt, but her disability interfered when she tried to transfer the number system into the concepts. She struggled to put commas in the correct places when she wrote numbers and sometimes she didn't notice the commas when she read numbers.

Zero was another hurdle. We decided that might be because zero is very hard to image, especially in comparison to concrete quantities such as 15. Double zero figures such as 20035 required considerable work before they fit into a meaningful image.

After many trials that included counting out and clapping the comma spacing (1-2-3 clap, 1-2-3 clap, etc.), Britt could describe the meaning of any numerical equivalent. We both knew she understood even complex versions of our number system when she could explain the value of an enormous number with a large decimal like 34,687,934.871.

"Okay," she said, "There is a thousands comma and a millions comma—and we have 3 numbers after the period. That means 34 millions comma, 687 thousands comma, 934." She paused, took a breath for courage, and continued, "After the decimal, numbers are getting smaller, so 8 is tenths, 7 is hundredths, and 1 is thousands. That means 871 thousandth."

Britt added, "I think it's neat that every number has a period even though you're only supposed to put it down when numbers come after it."

I blinked, trying to understand what she meant. Britt was recalling our review—4 really was 4 decimal point (4.) and the decimal point was not written unless you needed to include a number less than one that followed the decimal. She was reading and writing numbers accurately, only she still had to go back to basic levels and review the fundamentals as she deciphered the quantities. No wonder this process took her a bit longer than most students!

She smiled at my frown and finished her answer. "That means tenths, hundredths and thousandths. Okay," she summarized, "put together we have 34 million, 6 hundred and 87 thousand and 9 hundred and 34

> *Number position is a painful obstacle for anyone who does not perceive space accurately.*

and 8 hundred seventy one thousandth." We were ready for "x" and "y."

Britt had no difficulty substituting letters for numerals to create algebraic expressions. The concept of negative and positive numbers was much harder.

Britt looked at me with her blue eyes reflecting concern as she repeated, "Eight minus a minus two is x (8 - (-2) = x)." After a long pause she said, "That doesn't make any sense."

I had just explained the idea twice, once in my own words and once using a textbook. Yet the traditional number line found in algebra textbooks

Carol: Cracking Algebra's Code, continued

added to her confusion. The number line supposedly helps students understand negative numbers and to both add and subtract positive and negative amounts by counting to the right or left along the line.

One day I tipped the number line into a vertical position so that it read like a thermometer. The confusion disappeared. Within minutes, Britt was moving up and down the line, calculating the effect of negative addition and subtraction. By eliminating right-left confusion, a vertical number line allowed Britt to see the effect of adding and subtracting negative and positive numbers. The image of a thermometer helped Britt master a spatial process.

Britt explained her success, "Negative is down, like the temperature dropping. Positive is up like the sun warming the world up. Everyone knows that. And negative is below zero—really cold."

The algebraic sentences in the textbook included other pitfalls. For example, Britt read "seven more than twice a number is thirty-five" and wrote the phrase as $7 + 2x = 35$. The textbook authors and most other students wrote this as $2x + 7 = 35$. Using this equation to

> " Within minutes, Britt was moving up and down the line, calculating the effect of negative addition and subtraction. "

solve a problem could result in the correct answer, but this was not always true. The problem came when she dealt with a phrase such as "eight less than four times a number is twenty." This she wrote as $8 - 4x = 20$. The correct expression is $4x - 8 = 20$. In this case, the answers were not the same.

"But I wrote down what it said," Britt insisted.

She had, but the problem asked her to transpose the order, and she did not recognize the spatial demand. The problem was not conceptual to algebra; it was managing language to describe a spatial expression. Britt needed to image the event.

The phrase said she had $4x$ things. She imagined these as items in her hand. She realized that she was taking eight things away from that. She could then write the mathematical expression correctly. She had to realize that she had the $4x$ on hand before she could take eight away from it. Her route to understanding algebraic expressions was through creating specific and accurate images and reshaping the language to fit the image. Only after this first step did she make the translation to an algebraic expression.

The key turned out to be images and language woven together by creating and recreating illustrations from the textbook. Britt learned algebra, and I learned about rerouting and rebuilding her conceptual base for mathematics.

ಙ ಙ ಙ

The Source for Visual-Spatial Disorders   110   Copyright © 2002 LinguiSystems, Inc.

Britt: Life Beyond the Classroom

I studied a lot, but I didn't study all the time.

Every day, at 8 a.m. sharp, the entire upper school gathered in the stately dining room that overlooked Puget Sound for a mandatory morning meeting. At this time, announcements were read and important information relating to any class schedule changes were communicated. The student body president was in charge of reading the announcements and handing out schedules for sports teams, chapel services, language club meetings, and auditions for the school plays.

I arrived at school with just enough time to gather my books from my locker and find a seat in the dining room before the 8 o'clock bell. As I sat at one of the large, wooden tables waiting for announcements to start, I glanced over my list of vocabulary words I had been studying for the French quiz Madame Maxwell was planning to give that day. Then with a loud rap of her gavel, the student body president called the meeting to order.

"Okay, I have quite a few announcements to go over this morning, so let's get started," the president informed us.

"Anyone interested in playing volleyball or basketball this fall should sign up in the gym after school today. There will be soccer tryouts for the Stadium High team at 4 p.m. tomorrow, and anyone who wants to play soccer should see Ms. Adair." The president looked around the room and motioned to Ms. Adair, our P.E. director, and then returned to the sheet of announcements.

I tried to imagine playing volleyball or basketball and let out a little laugh. Being forced to play volleyball during P.E. was bad enough; there was no way I'd consider playing for fun. I immediately dismissed the idea of playing soccer, not wishing to relive my unhappy soccer experiences. Happy to stick to my schoolwork, I dismissed the

> **" Excited by the idea of being on stage, I marked down Friday's audition in my notebook. "**

athletic announcements and thought about the mythology reading for Civilization class.

"Mr. Selfe will hold auditions for the fall play on Friday in the new Kemper Theater. This will be the first production to take place in our new auditorium," the president said as Mr. Selfe stood up to explain the process in greater detail.

"This year we are going to put on a 1900s-style variety show. The main program will be a scripted melodrama. However, I'm also looking for people with talents like tap-dancing and singing to fill in between the acts of the play. So, if you're interested, come and see me on Friday afternoon." Mr. Selfe sat down after he had finished his speech.

"Now, that's for me!" I thought. Excited by the idea of being on stage, I marked down Friday's audition in my notebook. Our meeting adjourned with another rap of the gavel. As I made my way to class, I wondered whether I should have anything prepared for the audition.

Friday afternoon came and I headed down to the theater, wondering what would happen. Mr. Selfe sat in the house seats and handed out scripts. We were sent up on the stage and assigned roles to read. Every once in a while, Mr. Selfe would direct us to switch roles and read another scene. After everyone had finished reading, Mr. Selfe thanked us and announced that he would decide over the weekend. The final results would be posted on the door of the front office Monday morning.

On Monday, eager to see if I had a part, I hurried to the front office. My heart raced as I read down the list and found my

The Source for Visual-Spatial Disorders

Britt: Life Beyond the Classroom, continued

name next to the title "Page girl." "What kind of part is that?" I wondered to myself. Of course, I had envisioned playing one of the main characters in the melodrama and felt a little let down. "The first rehearsal will be held this Thursday afternoon" was printed at the top of the cast list in bold letters. "Well, I guess I'll find out more then," I thought to myself.

At rehearsal, I found that my part entailed walking across the stage holding signs that introduced the different acts in the play. My role in the play tied all of the different parts of the show together. I didn't have lines to learn; I simply had to appear on stage and prepare the audience for the next act. As rehearsals progressed, I experimented with some different entrances and figured out what worked and what didn't. By opening night, I had a routine that made everyone laugh.

With at least a third of the upper school student body and many of the teachers involved in the production, I gained a wonderful sense of community and made several friends. I relished being on stage during rehearsals and chatting in the green room between acts. I continued with drama during high school and enjoyed the balance it brought to schoolwork, providing a small break from the worries and insecurities of homework and classes.

In addition, I joined French Club. Meetings were held once a month during lunchtime, so French Club acted as a natural extension of French class. Aside from the monthly lunch meetings, we gathered several times a year to watch movies or to prepare a French meal and practice our conversation skills.

Once a year, the French Club helped prepare for International Day. Everyone at school looked forward to the special lunch and variety show. Both the French Club and the Spanish Club, along with the international students, prepared foods that represented different cultural traditions. As I stood in the lunch line, intriguing aromas wafted from the dining room and lured everyone in for a closer look at the menu. The choices ranged from *croque misures*, *croissants*, *brie*, *baguettes*, *pain au chocolate*, and *café au lait* prepared by the French Club along with the hot chocolate, *curros*, *tamales*, and *pollo con riso* made by the Spanish Club. The international students contributed an array of different dishes from *Miso* soup, teriyaki chicken, and tempura to shepard's pie and trifle. After lunch there were songs, plays, and skits performed in different languages with translations.

As an Episcopal school, Annie Wright required us to attend chapel three days a week. At first I understood chapel services as an obligation required by school tradition. The chaplain, Father Berge, oversaw the day-to-day administration of the chapel and taught several courses. Sometime during the year, I began to see chapel as the center of the school and an integral part of our education. At this point, I joined Raynor Guild, a group of students and faculty that tended to the chapel as a way to give back to the school. We decorated for the "Lessons and Carols" service during Advent, lit the candles during chapel services, stripped the chapel for Lent, and took care of the flowers as well as the brass polishing for special occasions such as Christmas or graduation.

The time outside of class became as important to me as the time in class.

> **" The time outside of class became as important to me as the time in class. "**

෴ ෴ ෴

Julie: Teaching Britt to Drive

During her sophomore year, Britt turned fifteen and a half, old enough to obtain a learner's permit to drive. I mentioned driving to Britt, but she said she was too busy with school to think about it.

I remembered when I got my learner's permit and how excited I was about learning to drive. Clearly Britt did not share this excitement. So I put it aside, figuring the topic would come up again. When school was out in June, Britt reluctantly got her learner's permit and signed up for a driver's education class offered by the local public high school.

When I signed the papers for the permit, I did not realize that Britt would have trouble learning to drive. She was certainly mature enough; she had exhibited good judgment about all sorts of things, and she was reliable and careful. I did not think that she would have any real difficulty taking on the job of driving a car.

Even though she had the permit, Britt wanted some practice before she actually started the course. So one Sunday afternoon, we drove to a deserted parking lot a few miles from home for a practice session. Britt had always been comfortable around the car; she had occasionally retrieved keys from the ignition or opened the trunk. But today, as Britt took the driver's seat, she acted as if she had never been in a car before. I suggested she adjust the rearview mirror, but she looked around as if she didn't know what I was talking about. I put my hand on it and then she did the same, but she asked what she was supposed to see. "Adjust it so you can see what's behind you," I said.

"Oh, I see," Britt said, reaching up to adjust the mirror. As we went through the other adjustments, I felt as if Britt had never even thought about driving before she got behind the wheel. She hadn't noticed the brake or the side mirrors.

In the parking lot, Britt started out at a snail's pace, and we crept along crossing empty parking spaces. "Give it a little more gas," I suggested. Britt took me literally, and the car shot ahead, crossing a dozen parking spaces in an instant, and headed for the azalea bushes coming up just ahead.

"Not so much," I screamed, scaring Britt and myself. She slammed on the brakes which threw us both into our seat belts. Then I suggested that Britt drive in the lanes of the parking lot instead of across the marked spaces.

"What parking spaces?" she asked. She didn't seem to comprehend, so I explained. Once I described them, she saw them.

After about half an hour of practice, she was able to keep the speed even, and I decided that we should drive home. The road to our house was a two-lane country road with gravel shoulders on each side and little traffic. I didn't think there would be a problem. We headed out of the parking lot

> **"** *The car shot ahead, crossing a dozen parking spaces in an instant, and headed for the azalea bushes coming up just ahead.* **"**

slowly. Britt didn't seem to notice the shoulder as she drifted onto it and then back on to the road. I thought the "drift" was caused by her inexperience. Then, she drifted into the oncoming lane.

"You're in the oncoming lane," I screamed again.

"Mother, you don't have to yell," Britt said.

"But you were headed into oncoming traffic. What if a car was coming?" I exclaimed.

Britt said, "I didn't notice."

As we rounded the corner into our development, I instructed

The Source for Visual-Spatial Disorders

Julie: Teaching Britt to Drive, *continued*

Britt to signal the turn and then slow down. Britt signaled for a left turn instead of the right, turned on the headlights and the windshield washers, and made a hard right turn while hitting the accelerator. The car came to a stop on the beauty bark in the neighbor's yard, with the window washers still on high spraying soap and water. Britt didn't seem to realize how close she came to real danger. "Hmmm," she said, "I guess we're in the Soames' yard."

"Britt," I said, "I know you're trying. But I think I'd better drive the rest of the way home."

That night I wondered why Britt had had such a hard time with driving. I also wondered how I could allow her to drive when she seemed so oblivious to the road and the dangers. She would surely kill herself and someone else with her current level of understanding. On the other hand, how could she live a normal life if she couldn't drive? We had no trains and the nearest bus stop was roughly eight miles away. I wondered if this problem was in some way connected to her learning disability.

The next morning I called Carol to see what she thought of this puzzling situation. I described the events in detail. "She doesn't seem to notice lanes or shoulders, she didn't know what taillights meant, or what the consequences would be if you didn't slow or stop

> **" Britt didn't seem to realize how close she came to real danger. "**

before you got to the car ahead of you. She doesn't seem to remember where the controls are on the car. If you say, 'Turn on wipers,' you might get the turn signal. If she remembers to turn on the signal, she might signal right and turn left or vice versa. She is a disaster waiting to happen. I am really worried. This is so dangerous. She's signed up for Driver's Ed, but I don't think she can do it—at least not safely."

Carol listened carefully and then said she'd have to think about it and get back to me in a few days. Three days later, she called to tell me she thought it was part of the learning disability, and she wanted to spend a little time with Britt. I was more than happy to set up the appointment.

ಙ ಙ ಙ

Britt: Learning to Drive

Almost everyone has a story about learning to drive. I'm not sure how mine compares to others. I had never given driving much thought before I turned fifteen and a half. Frankly, I didn't care one way or another about driving, but I was expected to learn whether I wanted to or not.

I started Driver's Education with a sour attitude, and the class lived up to my expectations. Convinced that I would never be a good driver, I anticipated the worst from the class. My view of driving had already been tainted by the practice drive with Mom in an empty parking lot. I tried my hardest to get the feel of the car and the assortment of controls in front of me. Expected to know exactly how all of the buttons, levers, and knobs operated, I froze. Clumsily, I tried all of the levers looking for the turn signal. As Mom became frustrated, I became more nervous and confused. Trying my best did not seem like enough. Mom kept yelling out unintelligible directions, and I tried desperately to keep the wheel straight while searching for the turn signal.

This experience left me rattled and discouraged. I knew that I loathed driving as well as everything else related to cars.

Most of Driver's Education was held in the classroom where we discussed various points of driving and rules of the road. There were 20 other students in the class. We were divided into smaller groups for on-road practice. Except for

> **"** *As Mom became frustrated, I became more nervous and confused.* **"**

me, everyone was eager to get out on the road.

After more practice with Mom, she gave up trying to teach me to drive. In pure frustration, Mom handed the responsibility over to my stepdad, who took on the challenge with gusto. At stop signs, he made me stop exactly at the white line. He looked out the window to make sure the tires were an inch behind the line.

One Saturday, he took my stepsister Sarah with us to practice parallel parking with orange road cones in an empty parking lot. She had been promised ice cream when I was successful, so Sarah sat in the back of the car, cheering when I had successfully parked the car between the cones.

After our practice with parallel parking, my stepdad decided it was time to try the freeway. This seemed like a great leap to me, and I blurted, "I'm not ready yet."

"Don't be a wimp, Britt!" Sarah and my stepdad said at the same time.

Somehow I made it onto the freeway, going north to Seattle in what, fortunately, was very light traffic. "Just remember to check your mirrors, keep up your speed, and whatever you do, Britt, don't stop," warned my stepdad.

My luck gave out as a sea of red taillights spread across the freeway ahead. I kept going.

My stepdad said, "Slow down."

I lightly touched the brake, not realizing why he wanted me to slow down.

"I said 'SLOW DOWN,'" yelled my stepdad.

I tapped the brakes again, thinking I had slowed down enough.

"Stop!" he yelled as he stepped across the center of the car to put his foot on the brake on top of mine. The car skidded to a stop just inches from the car ahead of us.

"Why didn't you stop when I told you to!" he said, anger clipping the words to sharp points.

"I did what you told me to. I slowed down. Next time, don't make me go on the freeway when I tell you I'm not ready," I retorted. Angry that I was

Britt: Learning to Drive, continued

being blamed for something I had no control over, I added, "You said, 'do not stop on the freeway.' Besides, anyone who stops on the freeway deserves to be hit." After that I don't remember the trip home.

Continued frustration with driving made it apparent that I needed a different way of learning to drive. That led to a discussion with Carol about driving and its relation to spatial abilities. First of all, we measured the width and length of the car in paces and arm lengths and talked through the measurements. Just like the ink we had used to measure an inch, I now had a concrete, physical understanding of the car relative to the road.

We took many gradual steps that I described in my own words until I felt I understood the action. This method was slow, but it helped me know about the car and the steps I needed to take to drive.

The other part of Driver's Education was on-the-road training and proved to be the most difficult part of the course. Our instructor favored some members of the class, continuously praising their efforts and constantly criticizing others, making them feel inadequate. At first, I believed it was only spatial difficulties and inability to judge distances that sparked the teacher's derogatory comments and harsh criticisms about my driving abilities. One driving session, I caught the eye of another girl in the car and saw her frustration as he sharply criticized her. Later I saw tears in her eyes, and she mouthed a few words to me, validating my view of the teacher. I nodded back in agreement. His negative approach had the effect of making me terribly nervous once I got behind the wheel.

"Your driving hasn't improved much and you need to practice more," the teacher said to me toward the end of the class. "I don't think you're ready to have a license."

"I have practiced, probably more than most of the students in class, and I'm doing my best," I said angrily.

I tried hard and managed to finish the course with Carol's help. Diligently, I read the chapters in the textbook and passed the tests. Eventually, I passed all of the on-road tests as well, including parallel parking.

In late August, I turned sixteen. I had decided several weeks before that I wanted to take the driver's licensing test. Some of my other friends already had their licenses, and I did not want to be left behind. The appointment at the Department of Motor Vehicles (DMV) had been arranged weeks before my birthday, and I wanted to get the test over with. On the day of the test, Mom and I went to the auto shop to pick up our car that was in for repair. The mechanic had promised that the car would be ready by 1 p.m. But when we arrived at the shop, the mechanic said the car would not be ready until the next day.

With the long wait at the DMV for an appointment, we decided that I would have to take the driver's test in the rental car. Although the car was the same make and model as our car, the controls were completely different. Mom drove to the appointment, pointing out and describing the differences between the two cars so that I would be ready for the DMV examiner.

Nervous, I took a few moments to acquaint myself with the controls in the rental car. The gearshift was located on the side of the steering wheel instead of between the two front seats. I had to get used to the fact that it had one long front seat, and I could not reach

> " Continued frustration with driving made it apparent that I needed a different way of learning to drive. "

The Source for Visual-Spatial Disorders

the pedals as well as in our car. The DMV examiner got into the passenger seat with clipboard in tow. I was so nervous. She began barking orders: start the car, turn right, turn left. At that point, I wanted to forget the whole thing. Then we approached a four-way stop, and I came to a full stop. I waited too long, letting an extra car go ahead of me when I had the right-of-way. The officer yelled, "Stop," and told me to pull over, explaining that I had failed the driving test.

With quivering bottom lip, I drove back to the DMV. I didn't cry until Mom got into the car. I had had it with driving and didn't care if I ever got a license.

In mid-September, I got back behind the wheel, practiced, and regained my confidence.

By October, I was ready to try again. This time I had the car I was used to. When the examiner got into the car, he saw that I was nervous and said, "You look nervous, but I'm the one who should be nervous. You have nothing to worry about." Politely, he gave me directions which I followed perfectly. When we pulled into the DMV, he told me that I had done a good job and handed me the paperwork. I had a driver's license.

The next problem with driving came when I had to learn my way around. It's easy to get lost, especially if you don't know the difference between east, west, north, or south.

> **"** *It's easy to get lost, especially if you don't know the difference between east, west, north, or south.* **"**

೧ ೧ ೧

Carol: Driving Is Spatial

Of course Britt would need to learn to drive. Julie and I agreed that this was important. We also agreed that the process of teaching her included more steps—unknown steps—than did usual driving instruction. Of course, we needed to be absolutely sure that she was safe on the highway and that everyone else was safe with her behind the wheel. But the question never was, "Could she learn?"; the question was, "How could she learn?"

I thought about previous experiences with several of the students in the clinic and with my own sons. All of them had to become aware of the sequences and actions to make the car do their bidding. And they needed to know the rules of the road and the common-sense additions to those rules: roads are slippery after it rains, especially if it has not rained for several weeks; it was better to be safe than right. Teenagers differ widely in their awareness of traffic and the ease with which they learn to drive. As I reminded my sons when they learned to drive, the accident rate for teens was twice that of any other age group.

"What was unique about Britt's learning to drive?" Julie had asked this question after an uncomfortable first attempt in an empty parking lot. I made a list to share with her. Every item on the list centered on space. A successful driver must perceive the width of the road and where the car was located in relationship to the center line and the edge. Britt needed to know how much space the car occupied, how much space other cars occupied, and to bear this information in mind while moving at different rates. She needed to feel how far a car will swing out when it turned a corner, how much space she would need to park, and how close she was to another vehicle and the appropriateness of that spacing at any given speed.

The first task, then, was to build an internal image for the dimension of her vehicle, other vehicles, and the road.

Parallel with learning about the car and road was building an image for the rules of the road. It was not enough for Britt to merely pass the written driver's test; she had to build images of herself following the rules. For example, she might correctly answer a question about when she could legally pass another car. That was not the same as judging the distance to allow sufficient clear space to return to her own lane before the approach of the oncoming car.

Julie asked a thoughtful question. "How will we know when she really knows the system of driving?"

"I guess it's the same way we know she knows anything else. She can put it into her own language," I replied. "'Verbally mediated' applies to driving just as much as it does to Algebra or American History."

Julie with her typical, practical approach said, "This sounds like it will take time. Let's get started." We set a planning time for the next week.

I tried to be conscious of each action I took behind the wheel that week. The runner cutting across the highway or the line of bicycles in the right lane were vivid examples of what Britt needed to anticipate when she was driving on the road. I slowed automatically to let a rapidly approaching car cut in front of me and thought, "How can Britt recognize the need to do that?"

She must allow for all kinds of other drivers, I noted, as I allowed a horse and buggy decked out for some special

> **"** It was not enough for Britt to merely pass the written driver's test; she had to build images of herself following the rules. **"**

The Source for Visual-Spatial Disorders

advertising event to have the right-of-way. Britt even had to be ready for the bizarre.

Confused drivers are out there too, I noted as the woman ahead of me turned right despite the fact that she had turned on her left blinker. And then there is the unusual, I thought the next day, as I braked quickly to allow a flatbed truck carrying a tank to make a wide turn.

In addition to the bizarre, confused, and unusual, Britt had to allow for the rash or negligent driver. The lady driving with a dog on her lap, the man with several children moving around inside the car, or simply the speeder on bad roads. Britt needed to build images for what should be and what could be.

Hearkening back to her finger dipped in ink to perceive one inch, I realized she needed to feel the dimension of the car by building an image based on her body size. Julie and I talked and planned. Perhaps pacing around the car will be enough, perhaps not. Perhaps she will need to know the dimension of the hood in hand-widths. We agreed that Britt needed to pace the width of the road and to translate that information into oral descriptions. Julie, superb at noticing vague or imprecise language, would probe Britt's descriptions for each experience. She would be absolutely certain that Britt knew the size of the car and road in relation to her own body.

Julie called, tense anxiety stretching her voice to flatness, to relate the freeway incident when John stopped the car after Britt had only slowed. Guilt swept over me as I thought about the accident that could have happened.

"Well," I said finally, "there is something else we haven't talked about or taught Britt. Absolutes! I think Britt was very fortunate."

Julie's voice remained apprehensive and now was also very quiet, "What do you mean, absolutes?"

"Think about living with a spatial disability, Julie. You would crave, even demand, absolutes as something you could count on. If told something was an 'always' or a 'never,' you would fiercely latch onto this assurance. You just said Britt had been told 'never to stop on the freeway.' For Britt that meant absolutely never. Of course she understood that applied to others. No one stops on the freeway. The concept of stopping on the freeway was outside her frame of reference. Besides," I added, "you said John told her to slow down, and she did just that. We have to help Britt understand that judgment is the name of the game."

Julie asked, "But how could she believe that those cars deserved to be hit?"

"That's another problem with absolutes—consequences are tied to the absolute. So if you break an 'always' or a 'never,' consequences follow. The consequence is linked to the one who breaks the rule and that was the driver who stopped. Britt did not extend the idea to realize that if the car stopped and she struck it, injury would come in her direction as well," I explained and added, "We have more work to do." I heard Julie sigh before we said goodbye.

As I put down the phone, I started thinking about the driver's training manual. The manual is a set of information

> *" Guilt swept over me as I thought about the accident that could have happened. "*

in print and drawings. Britt needed to translate these into three-dimensional space.

When the page said, "Do not park within fifteen feet of a fire hydrant," Britt needed to feel fifteen feet. Furthermore, she needed to anticipate the placement of fire hydrants.

A separate part of the task, then, was translating the words and drawings of the manual

Carol: Driving Is Spatial, continued

into three-dimensional reality. Then Britt needed to put that reality into her own words. One way to start was by writing her own test questions. Her needs went beyond answering the questions posed by the official test.

Understanding the rules and driving safely was one thing. Knowing where she was, was a separate issue. I remembered how difficult it was for Britt to adjust to new locations. She had been lost several times at each new school. Once when visiting her father, she had been so lost she phoned for directions only to discover she was one block from her usual route. Driving would allow Britt to travel sufficient distance to get utterly lost. So the problems of mapping her environment and knowing where she was in relation to where she wished to be now took on an entirely new dimension.

Britt's visual memory for landmarks had proven accurate and reliable. She could remember them and she could describe them. The time had come to create a map of her territory with the landmarks selected by herself. She would need landmarks to orient herself wherever she was and in whatever direction she was traveling. This could also build her confidence as she gained greater control of her environment.

ଔ ଔ ଔ

Julie: Commuting with Britt

During Britt's freshman year in high school, my husband began to feel the need to move closer to his work. One winter morning when the weather was particularly bad, he had spent three hours in the morning commute. When he finally did get to work, he called to say he was never coming home again. He did finally come home three days later when the ice melted, but it was clear that the commute was a serious problem for him. After some lengthy discussions, we decided that we needed to move. For Britt and me, the move meant that we would now have a 55-minute instead of a 20-minute commute. Britt solidified the plan when she said, "I'm not changing schools."

Though we had all agreed that the move was the right thing to do, it was difficult to be up at 6 and on the road before 7 a.m. Britt and I had been feeling grumpy about our drive so early in the morning. After the first week, Britt and I decided to stop at McDonald's. For some reason, the Egg McMuffin and coffee seemed to help our moods. We chatted about everything from my childhood to plans for Thanksgiving.

One afternoon in October, Britt was unusually quiet when I picked her up. I asked her what was wrong.

"Oh, I have a big history test on Friday, and I'm worried about what Ms. Soucey will ask," she replied.

"What are you studying?" I asked.

"The early American period," Britt replied without enthusiasm. "Ms. Soucey's tests are so hard; everyone is talking about them." I could see tears begin to pool in Britt's eyes.

"Tell me about the First Settlement. It's been years since I studied that."

Britt started talking. I listened carefully and asked questions. It became clear to me that Britt knew a lot more than she thought she did.

The next morning, we climbed into the car and Britt asked if we could continue the discussion of history. Britt's tears

> **" *She bounded into the car and without taking a breath began to talk.* "**

were gone and she talked enthusiastically about people and consequences.

That afternoon I picked Britt up at the usual time. She bounded into the car and without taking a breath began to talk. "We reviewed for the test and guess what?" She continued without giving me a chance to reply. "Ms. Soucey asked a lot of the same questions we talked about in the car this morning."

"Did you answer in class?"

"Yes, and I think Ms. Soucey was really surprised at how much I knew. I raised my hand for almost every question she asked," Britt replied.

"So what did you talk about in class?"

Britt described the class discussion and raised questions she was still pondering. She pulled her American history book from her book bag and began searching for an answer to a question.

"Here it is," Britt exclaimed.

Britt continued to talk about Jamestown, the Puritans, and the Triangle Trade. When the talk slowed, I asked her another question. Before we knew it, we were pulling into the driveway.

The next morning in the car, I asked Britt if she would like me to ask her more questions for the history test. "Yes," she said. "But this time I'm going to tell you what to ask."

"Ask me about the taxes the British imposed on the

The Source for Visual-Spatial Disorders

Julie: Commuting with Britt, *continued*

colonists." It seemed that by hearing the question aloud, Britt was better able to focus her answers and remember the information. I knew Ms. Soucey gave essay tests, so I pressed Britt for relevant detail.

> **"** *It seemed that by hearing the question aloud, Britt was better able to focus her answers and remember the information.* **"**

As she and I talked, I suggested she think about the generalizations she was making and how she would back those up with details, examples, or facts.

As we pulled up to the school, Britt said, "Okay, I'm ready. I don't want to talk about the history test anymore."

These first encounters in the car made me realize that the time provided opportunities for all kinds of learning. Britt spent so much time on the road she had little time for reading, so I ordered versions of her literature books on tape when they were available. If they weren't, I ordered another book by the same author or a tape about the author. The taped books not only helped Britt get through her reading, they also gave us topics for conversation, and we often stopped the tape to discuss a topic or a passage in more detail.

Most days we used the time in the car for study, but sometimes we were too tired. On those days, we turned on the radio and let ourselves start to daydream.

We even wrote papers in the car. One day Britt jumped in and said, almost before she closed the door, "Ms. Soucey gave us the best paper topic today."

"What is it?" I asked.

"If we could reinvent holidays, what would we invent and why? Right now, we have Labor Day, Thanksgiving, Christmas.... But who says these are the best holidays? Maybe there could be new holidays that better reflect our values."

"What an interesting assignment," I said. "How many holidays can you create?"

"Seven. I think I'll keep Thanksgiving because I like turkey and the idea of being thankful," Britt said. "Because I'm a Christian, I want to keep Christmas, and because it's fun, Fourth of July too. That's the best holiday, and it is the birth of our country, but I have four more to figure out."

By the time we pulled into the driveway, Britt had decided on four more holidays and had a tentative outline for her paper.

The commute we had both dreaded turned out to be a good time for both of us. We talked over problems and listened to and discussed books. It turned out that Britt did her best studying this way because she was using language to learn.

ಙ ಙ ಙ

The Source for Visual-Spatial Disorders

Britt: On the Road

On most mornings I was barely awake enough to throw on my uniform, eat a bowl of cereal, and grab my backpack before jumping into the car. We left the house at 7 a.m. Winter days in Seattle are short and the sun rises at about 8 o'clock and begins to set by 3:30. During the winter, Mom and I set out on our journey south to Tacoma well before sunrise and watched the sky turn a light luminescent gray as the sun started to glimmer in the east. I looked forward to the mornings we had "Breakfast Club." We listened to the local news radio station and slowly ate breakfast and sipped coffee. My brother who was going to school in Seattle was always jealous of our breakfasts out.

Many mornings we listened to the books on tape that Mom had ordered. As we pulled onto the freeway, I would start the tape where we had left off the previous afternoon. Mom had not been able to find *The Great Gatsby* on tape, so she ordered *Zelda*, a biography of F. Scott Fitzgerald's wife that detailed their life together.

We always had discussions about the books, and sometimes I stopped the tape to ask Mom questions. "So she's saying Fitzgerald used Zelda's journals for his own books? I wonder if he kept her from publishing her work? I don't believe she was really crazy; he made her that way," I said, keeping a finger on the pause button.

"Well, Britt, that may be true, but I think you're jumping to conclusions. But let's keep listening and see if we can find out," Mom replied.

Other mornings we talked about the quizzes or tests I would be taking that day. We practiced French vocabulary and conjugations. I decided that Mom needed to learn French. I threw out vocabulary words and helped her with her accent.

On the way home, we often discussed History and English paper assignments. We brainstormed paper topics and usually came up with one brilliant idea. Then talking through the different sections of the paper usually helped me figure out the sequence of ideas, such as why a certain section should come before or after others. Once I got to my desk to write, it became easier to organize and sort through the bits of information I had gathered for the paper. Talking about paper topics helped me understand what I already knew and what I needed to clarify. Frequently I discovered that I knew much more than I thought I did.

Some mornings time passed quickly, and I wanted to continue our conversation; other times, the drive was simply tedious. Once in a while, I'd

> " *Talking about paper topics helped me understand what I already knew and what I needed to clarify.* "

fall asleep. I was exhausted after a full day of classes that were intense and required so much concentration.

Although I spent about ten hours a week in the car, I didn't mind the commute because I thought more thoroughly about paper topics and test material than I would have otherwise. Mom was a captive audience; she liked helping me and asking me questions that helped me think about whatever I was studying from a different perspective. Her questions were very useful not only for high school, but for preparing me for college.

Britt: Geometric Shapes

As I picked up my class schedule, I glanced over the list of classes I'd be taking: French III, World History, World Literature, Art History, Biology, Swimming, and Geometry. Oh no, I thought to myself, I'm not sure I'm ready for another class with Ms. Brown. I paled when I thought about geometry and the humiliation of failing algebra. But what choice did I have?

I was determined to give geometry a chance. For the first week of class, I followed all of the assignments and understood the basics of geometry. I had no trouble naming the shapes and identifying the angles. Ms. Brown chalked the problems from the book onto the board—rectangles, squares, isosceles triangles, equilateral triangles, and right triangles.

I tried to follow as Ms. Brown explained the theorems that accompanied the shapes. At home, I diligently trudged through the pages of homework, trying to follow the instructions in the book. This only confused me and muddled Ms. Brown's explanation even further. I felt my grip on the material quickly slipping away.

Two weeks into the semester, Ms. Brown handed out the first major geometry exam. A wave of apprehension took hold. I tried to calm down, reminding myself that I had spent at least three hours reviewing for the test. Ms. Brown told us to turn over our exams and begin. I recognized the steps I needed to solve the first problem and went to work, plugging in numbers, multiplying or dividing, and entering the answer on the test page.

After solving the first section, I moved on to the next portion of the exam and encountered a row of questions that appeared miles from anything we had covered in class or in the textbook. I looked up at the clock; time was passing quickly.

Holding back a flood of tears, I bit my lower lip and forged ahead. I applied theorems that seemed appropriate and hoped for the best. It was hard to tell if the theorems didn't work or whether I had chosen the wrong theorem. Both were highly possible.

For every question, I made several attempts at finding the right solution, and the paper became thin and smudged where I had tried to erase multiple mistakes.

The hour ended and I had finished two thirds of the test. I set my pencil on my desk and handed in the test, trying to contain my disappointment.

> *" I set my pencil on my desk and handed in the test, trying to contain my disappointment. "*

By the end of my third week, I needed Carol's help. Even though we had done a bit of work in summer to prepare me for geometry, I clearly was not ready. I went to see Carol and explained the situation. Carol tried to supplement the classroom lessons by putting the theorems into language. We built several models using toothpicks and colored 3 x 5 cards to explain the theorems.

Time became an issue. It took longer for me to finish the steps than the teacher would allow, and it was clear that a new approach was needed.

Mom arranged a meeting at school with Carol; Ms. Soucey, my advisor; the head of Upper School; Ms. Brown, the math teacher; and my aunt Jill. Even the biology teacher, Mr. Timson, attended the meeting. When I walked into the room, I felt like a specimen under a microscope.

I was questioned about my current situation in geometry, my past work with Carol, and then, finally, what I'd be able to gain from working on geometry with Carol instead of staying in Ms. Brown's class. I whipped out the models Carol and I had constructed and showed each example of "Britt geometry."

While I listened, Mom and Carol made the case for geometry at the clinic. The head mistress stood up, signaling the end of the meeting. Mom, Carol, and I left, giving the teachers a chance to deliberate my case. When all of the options were weighed, they approved setting up a geometry class with Carol at the clinic.

Thankful that I had been released from Ms. Brown's geometry class, I was overjoyed to be working with Carol. Carol set up a syllabus that covered everything I'd be expected to learn in geometry. The first matter of business was building a logic base. Carol pulled out a logic book used at a local high school to introduce basic rules of logic. At first this seemed strange, but I soon realized that this exercise helped me understand the logic of geometric argument.

We built shapes and described them. I started with cardboard and scissors and once I had constructed the shapes, I labeled them and discussed their applications. We used cardboard, string, Styrofoam cones, paper clips, clay, 3 x 5 cards cut into shapes, and even a beach ball. We made the shapes and used colored ink to add definition to the postulate.

We then put language with the shapes. Before we could use the language from the book, I had to describe the concepts in my

> **" *We made the shapes and used colored ink to add definition to the postulate.* "**

own words. For most theorems, we went through several different steps.

In Ms. Brown's class, we were assigned countless problems I did not understand. Doing a hundred problems would not have instilled the concepts. But with Carol, I built one model that gave me a tangible basis to solve similar problems.

ಙ ಙ ಙ

Julie: Making Geometry Happen

At Annie Wright, the new year began with a potluck at the end of the first week of classes. Hamburger casseroles and green Jell-O salads filled the long tables, and students and parents filled the front lawn of the school. The students found friends, and the parents searched the crowd to locate their daughter's teachers.

Cookies in hand, I stopped to gaze at the crowd and to find someone who could tell me where the desserts should go. Britt was tugging at my arm. "There's Ms. Soucey," she said, "and Auntie Jill, I mean Ms. Jueling. Come on, don't you want to say hello to them?"

"Just a minute," I said as I handed my cookies to the father who was in charge of desserts. "I also want to meet your geometry teacher."

I chatted with both my sister, Ms. Jueling, and Ms. Soucey about how much Britt liked Annie Wright, Art History, and History. Even in the first week of classes, they had made the subject matter come alive for Britt. I continued to look for the geometry teacher, but I was caught in the crowd moving toward the food tables. After dinner, I saw Ms. Brown standing alone near the front door to the school. I walked up to her, my hand out, a smile on my face. Unenthusiastically, she shook my hand as I reintroduced myself as Britt's mother. She hadn't been fond of Britt when she was a freshman because she saw Britt

> ❝ *Britt's anger, frustration, and hurt poured from her heart to her lips.* ❞

as lazy and inattentive and her opinion hadn't changed. I explained that Britt was still struggling with math, but she was working hard, hoping to succeed. Ms. Brown replied that Britt seemed to be doing fine so far but that she needed to focus more on her work.

Although Ms. Brown did not have the enthusiasm for Britt's learning that the other two teachers had, I didn't feel discouraged yet. Every year, I hoped that the work we had done with Britt the year before would be enough to carry her through the new and difficult tasks that lay ahead in the increasingly difficult curriculum. I left the picnic thinking that the year was off to a good start.

The euphoria of the first week of school passed, and Britt settled into the routine. Britt was working hard, studying as soon as she arrived home in the evening until bedtime and again in the car on the way to school. She spent an inordinate amount of time on math, but each day she completed the homework assignments and turned them in on time. To her dismay, she seldom got all the solutions right, but she continued undaunted. There had been a quiz, and she didn't do as well, but she was still passing with a "C."

Two weeks into the term, Britt took the first test in geometry. She had studied hard. The next day when I picked her up she looked very unhappy. Britt slammed the door, and as I pulled away, her eyes filled with tears.

"What's wrong, Britty?" I asked. Britt's tears turned to sobs as she explained that she had failed the geometry test, the only one in the class who had failed.

"And I worked so hard. I thought I knew it. And when Ms. Brown returned the tests, she thought I just hadn't studied. But I had. She doesn't understand at all." Britt's anger, frustration, and hurt poured out from her heart to her lips.

I assured her that I knew how hard she had studied for the test. "Britt, I know how bad you must feel. I feel bad for you. But there must be a way to deal with geometry. We knew going in it was spatial. You're smart enough, look how well you're doing in your other classes. There has to be a way

Julie: Making Geometry Happen, continued

to get around geometry, or through geometry. We aren't going to let you fail," I said.

"What can you do? I'm the one who has to take that class," Britt had started to cry again.

"Okay, I'm going to call Carol tomorrow and see what she says. Then if Carol thinks it's a good idea, I'm going to call the school and set up an appointment with the math teacher, the head of upper school, Carol, and whoever else I can think of. We need help solving this problem, and I'm going to see we get it."

I called Carol the next morning and explained the situation. She asked that I bring Britt in with her geometry book. By 3:30 that afternoon, Britt, Carol, the geometry book, and I were in Carol's office. Carol talked to Britt about the class, the concepts, the book, and the teacher. When they had finished, she asked me to call her at home that evening. When I called, Carol explained that she felt it was very unlikely that Britt could be successful in that class even with tutoring and regular help. "It's not that Britt can't learn it," Carol said. "She just can't learn it the way that teacher is teaching it."

"So what do we do?" I asked.

"Let me think about it," Carol said, "and I'll get back to you tomorrow afternoon. We have to do something right away; this is not a healthy situation for Britt."

Carol called shortly after 3 p.m. and suggested that Britt take the class at the clinic. She would learn the essential concepts through actually building the models. She would do 3-D geometry. But Carol was realistic about the problem; she explained that she didn't think the school had ever let a student take a class off site and get credit for it. We would be asking for an unusual accommodation—we wanted Britt to take her geometry from Carol at the clinic and get Annie Wright credit for it.

I called for the meeting.

I chose my clothing carefully— black business suit, crisp white blouse, red and black tie, black pumps, black briefcase. At this point, I believed that if Britt did not get special accommodations for geometry, she would

> **" *She needed to learn the concepts, but she had to learn them in a way that would work for her.* "**

not be able to graduate from high school, let alone go to college. This meeting might be my only opportunity to make the case for taking an alternative form of geometry. She needed to learn the concepts, but she had to learn them in a way that would work for her.

I called the learning disabilities specialist at the university and picked up copies of the federal law that guarantees that students with disabilities will not be discriminated against. I made several copies and tucked them in my briefcase. As one of my friends in the university English department said, "I was loaded for bear."

When I walked into Bishop's Lounge at Annie Wright, sunlight streamed through the leaded glass windows. I was surprised to see that all of Britt's teachers were there, including the counselor and the head of Upper School. Carol smiled at me as I took a seat. Ms. Brown, the last teacher to enter the room, chose a chair near the door.

The head of Upper School explained that I had asked for this meeting to discuss Britt's learning disability. I nodded and explained that I knew we all wanted Britt to succeed and that Carol Stockdale, a learning specialist who had worked extensively with Britt, might be able to shed some light on Britt's capabilities.

Carol spoke about her work with Britt, explaining how she learns and then focusing on the problems with math. As Carol

The Source for Visual-Spatial Disorders

Julie: Making Geometry Happen, continued

spoke, I watched the teachers. Ms. Jueling leaned forward in her chair listening carefully; Ms. Soucey and Mr. Timson were taking notes; Ms. Brown, the math teacher, had turned in her chair so she was almost facing the door. I wondered if she knew what her body language was conveying.

Carol proposed the geometry course she would design for Britt, saying that because of the timing, she had a sample of the curriculum rather than a complete course. She explained how important she felt geometry is, but Britt would have to learn another way. The head of Upper School looked interested, but said she would have to take it under consideration. What we were asking for was highly unusual. She turned to me for a final comment.

"Annie Wright is a wonderful school," I said, "because you take the needs of every student into account. I know you want to do the right thing for Britt. Sometimes it makes decisions easier if we're also doing what the laws encourage us to do. I have a copy of the law that protects the rights of students with learning disabilities. Do you have a copy?" I asked the headmistress.

> **" She explained how important she felt geometry is, but Britt would have to learn another way. "**

If I had been a dog, the hair on the back of my neck would have been standing up; I was ready for a fight. But instead of being antagonistic, the head of Upper School said she didn't have a copy, but would very much like to see one. I passed out copies to those who were still there. The math teacher had already left for another appointment.

"If you can approve the geometry course with Carol, we should start right away," I said. "I want Britt to learn the material, and I don't want her to get behind."

On Monday, the headmistress was on the phone—Britt could take geometry at the clinic with Carol.

☙ ☙ ☙

Carol: What's Your Angle?

The agreement with Annie Wright in the fall of 1988 gave sparkling promise to the fresh school year with opportunities unblemished by last year's struggles. At least those struggles had produced results. The agreement with the school was clear. No more Brown. No more unsympathetic teacher who refused to acknowledge Britt's needs. Instead I would teach Britt geometry as a Directed Study at the clinic. But what could and should I teach her? And how?

My planning folder grew thick from notes recording ideas and warnings to myself. Britt's school expected her to master geometry. So did I. But she needed some fundamental conditions in place before theorems and postulates would make any sense. How could she compare angles and infer relationships unless she understood the space they were measuring?

We had already established that systems could only make sense when she understood the foundation for that system. That included forms, shapes, measurement, and time. We had to build these foundations along the way. Of course we already had built some parts and pieces. Britt "knew" 12 feet was different from 12 square feet or 12 cubic feet. She understood this in the "real" three-dimensional world she walked around in. However, she could not routinely look at a two-dimensional diagram or picture and translate that to the real world of three dimensions. Some skills for this transfer, such as understanding point-of-view and overview, were not in place. The fact that many of these missing skills are never taught in school—are simply basic connections that people feel—made it harder. I might miss some of these. In fact, I was sure that I would. Perhaps my oversight would undermine Britt's progress.

Britt's cheery greeting cut through my worrying. We were meeting this warm September day to plan for our first semester's work. We had known each other for seven years since those first emotional sessions sorting out her learning needs. Of course, several of those years held only one or two visits. Now she was telling me how much she was looking forward to the year even though it meant "a weird schedule." I started to tell her my concerns and that this type of class hadn't been done before. She smiled, raised one eyebrow, and said, "So?"

We got to work. "You mean we do 'Britt geometry' and 'real geometry' all at the same time?" she summarized at the end of our hour of planning.

"That's it," I agreed, as we walked down the hall to meet the driver waiting to take her back to school. Her words made preparing the syllabus less intimidating. This was a class built on her strengths—her language strengths—not just the problems with space.

Later as I sat at my desk, I accepted that Britt fully trusted me to listen. And she would trust me with her confusion. I swallowed, wiped my eyes, reached for a new sheet of paper, and wrote "Directed Study in Geometry" across the top. We had begun.

The items on our "Must Do" list were gradually built into the class plan:

- Add the language of geometry to Britt's extensive vocabulary.
- Make sure Britt has the experiences to create accurate and specific images for the language. Images are foremost.
- Shift point-of-view. Besides images for geometry's language, Britt needs to flexibly shift point-of-view.
- Internalize time intervals. Mapping time needs to be in this geometry.
- Cover all basic concepts in a traditional geometry textbook.
- Learn traditional geometry sufficient to meet the demands of college boards.
- Create personal concept summaries for "Britt geometry" and "real geometry."

Carol: What's Your Angle?, continued

We began the class with stacks of graph paper, several geometry books, a high school level logic textbook, and various reference materials. Britt and I constructed examples of the postulates out of paper, tape, and paper clips. We started with postulates because they are "geometry talk" and to make certain that we included the basic terms and definitions in our image building.

We called our constructions "actions." Whole-to-parts was a very important approach. Therefore, we made 360-degree circles before we cut them up into parts and then reassembled them. We used cones, sticks, string, and other three-dimensional materials. Once she had made a construction, Britt *knew* the form and the associated postulates. We were creating perceptual bridges.

We also needed a format for translating each action into useful language. Transfer was our goal. We built images, and described in words. We described in "geometry-talk," which meant we created the symbolic form. Then we summarized the basic concept. Just as in that long-ago Constructions class, Britt built meaning through making objects and explaining the construction process and what each object illustrated.

Britt's notebook grew fat.

Once Britt had constructed a postulate from paper or string, the concept was obvious to her. Waving two small wedges Britt said, "If you add up the degrees of this angle with this other angle, you get the degrees in this united angle," she said, thrusting the two together and reaching for tape to stick them together for her notebook. I smiled as she labeled the "Angle Addition Postulate" and reached for a pair of scissors to make the "Supplement Postulate."

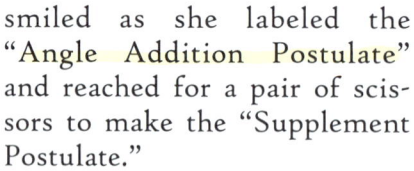

We read the logic book together, and Britt constructed statements following each form. During the same period, we created spatial design problems. Some of these were modeled from the exercises in the reference books. Britt learned to put the premise of the design into her own words. Then the solution was obvious to her.

"They set up a repeat of circle-square-diamond," said Britt.

"Then they omitted square," she added as she drew the missing link with a flourish.

As Britt advanced through traditional and non-traditional geometry, another concern edged its way into our sessions. Time is not a topic in geometry. But for Britt, perceiving time, understanding time, and managing time became a necessary survival skill.

Although we had fun during her geometry class at the clinic, Britt had no time in her daily schedule for play. Six challenging academic classes plus her geometry coupled with two hours of commuting each day made time a problem. Sequencing activities and making choices to plan her work was a nagging necessity. Over and over, the plans she made proved unrealistic. No, the hour after dinner is not adequate for reading American History, shaping a thesis sentence in Literature, and calling a friend. Time was very difficult for her to monitor. Deadlines were indeed hard to meet. But deadlines were even more stressful because Britt could not pace out the steps to complete major tasks with realistic time allotments. Real stress and perceived stress blended into one.

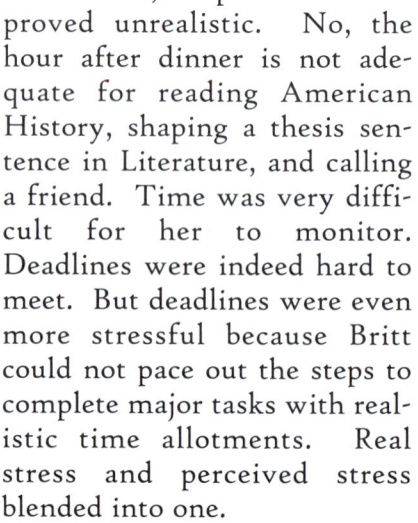

I mulled over the situation. Time is spatial. Britt has a spatial disability. Time is an interval

Angle Addition Postulate

Britt's Summary
- Action
- Language
- Symbolic Form
- Concept

The Source for Visual-Spatial Disorders

translated onto a clock-face system of numerical equivalents. Time is an abstraction and the clocks used to measure it rely on number relationships for meaning. Britt, with her spatial disability, could read time but she could not "feel" interval nor could she sequence activities realistically.

Plan sheets did not work. They just added another piece of paper to her work. Britt needed to "feel" time. We tried to use a class period as a measure for Britt to "feel" duration, but this just was not tangible enough for her to serve as a frame of reference. Sand timers were helpful but not portable into her day.

Britt needed an internal measuring stick. The commute proved to be the most "real" chunk of time. Perhaps this was tangible because the car moved through space as time passed. For whatever reason, Britt accurately perceived the hour of commuting time as a measurable interval. She could use it to estimate the duration of tasks. We used the commute hour for a base. A geometry assignment might take one and a half commutes. Studying for an Art History exam might take two commute chunks of time.

The next step was to extend this base to a 24-hour day. Her time plans for stressful projects allocated all 24 hours in a day and included sleeping as well as classes. Britt became better able to predict her time needs, although this remained challenging.

Britt also learned to use a control scale to hold stress at bay. She made a rating chart of *1* to

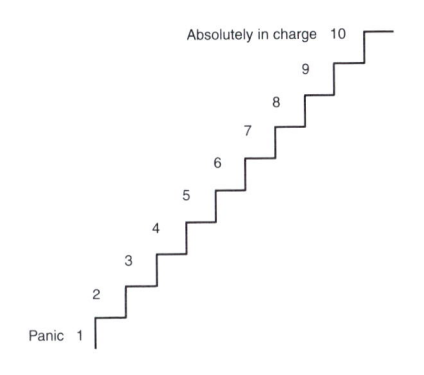

```
Absolutely in charge  10
                       9
                      8
                     7
                    6
                   5
                  4
                 3
                2
         Panic 1
```

10 with *1* being "panic" and *10* being "absolutely in charge." She identified and recorded feelings that went with each level. She also established the number she needed for a comfortable level of control. Realizing that she was at a 4 in Art History and wouldn't feel much better until she was at a 6 made the actions she needed to take much more apparent.

Britt and I worked our way through geometry, developing and adapting methods and theorems as we moved through crisp, fall days and into winter. She richly deserved the "A" grade she received at the end of the first semester. We continued to build and broaden each area as second semester progressed.

However, mapping was still a concern. This led to our messiest project of all. The idea seemed simple enough. Britt could build a broad overview by creating her own 3-D globe. Making the imaginary latitude and longitude lines visible could and should help her build a framework for understanding the location of continents and countries. But yarn for latitude and longitude slipped and slithered on the beach ball. Paint ran and dripped on the slick surface until continents smudged into each other.

Somehow, despite the mess, Britt described her discoveries. "Oslo was just below the 60-degree N. latitude line. The equator was at 0 degrees and it went through Africa and South America. Great Britain was really close to France—really, really close. And Africa stretched all the way down to Antarctica—well almost." Her images for countries became clearer and more personal with each description.

Britt was tactful. "Um," she said in her second effort to find

> *Britt accurately perceived the hour of commuting time as a measurable interval.*

Carol: What's Your Angle?, continued

words for the disaster, "I don't think this model is working very well."

She was right. "Interference due to 'technical problems,'" I said. We scrapped the globe.

The year held many more high points than low. Time became more tangible. Floor plans made overhead points-of-view more realistic. And angles, degrees, hypotenuse, and parallel lines assumed a natural reference in conversation that testified to meaningful images.

One spring day, Britt was describing a complicated history assignment from her favorite teacher. "She's asking us to shift our viewpoint 180 degrees from the author . . ." She stopped, looked up and laughed, "I think geometry just sneaked into my vocabulary."

I smiled back. "You can take that point-of-view, Britt," I replied. "But tell me, just what is your angle?"

We both laughed.

༂ ༂ ༂

Britt: Moving Forward

After surviving Civilization freshman year, my fears of Ms. Soucey and her class had been put aside. Freshman year had been the equivalent of academic boot camp and now that we had passed basic training, we were allowed to move up in rank. As sophomores we now had experience with paper writing, taking in-class notes, and taking exams. Although classes were still demanding, the element of the unknown was gone, and I focused on improving the techniques I'd already learned.

Ms. Soucey's sophomore year American History class set the course for the rest of my time at Annie Wright. I learned that history could be a voyage to another time. Ms. Soucey saw history as cultural, encompassing all aspects of a society, not just a timeline of events. History discussions always included a description of the culture and society surrounding important people and events. We were encouraged to take on different points-of-view and bring in concrete examples whenever possible.

By the end of my sophomore year, I had been able to catch on to the format for writing papers and had practiced developing arguments. I no longer worried about disorganized papers or exams. I still had trouble editing, but I felt that I knew what I was doing in terms of format and topic, and I could spend more time on my arguments and research.

The paper I wrote for American History at the end of my sophomore year on "Women and Education," in which I traced the development of education for women from the colonial period to the present, was the easiest and most interesting paper I wrote in high school. Mom and I conducted writing conferences in the car and at home after I had the ideas on paper. I learned the value of a writing conference, and I could see the difference in my papers. For the first time, I began to verbalize the outline of the paper and understand the organization and how transitions worked to create a link for the readers. I learned about point-of-view and how it played a part in writing a paper and constructing an argument.

As I waited for Mom to pick me up after school, I busied myself in the school library. I picked through the card catalog and made notes of any books I imagined might relate to my topic as well as those that would provide the background information I needed to balance the details. Pulling out armloads of books from the shelf, I piled them on the long, wooden library table in front of me.

As I sorted through the books, images of life in early America unfolded, and the women teaching their children to read were as real as if I had been in the same room. I slowly began to tie the roles that women played in education with the values and events of the time. I felt strangely energized as I made my discoveries and made connections between the bits of information I had gathered from the many books, hoping to find examples to build my argument and make the different histories come alive.

I used the book that Ms. Soucey assigned for our Civilizations class the year before as a guide for the outline, footnotes, and bibliography. By this time

> " *I learned the value of a writing conference, and I could see the difference in my papers.* "

everyone was familiar with the electric blue book, *The Research Paper Guide*. If anyone had any questions about outlines, footnotes, bibliographies, or punctuation, Ms. Soucey directed them to this book which we nicknamed the "blue bible."

I had a good feeling about the paper when I handed it in and

Britt: Moving Forward, continued

knew that it was better than anything I'd ever handed in before. I was confident that the conferences Mom and I had during the numerous trips to and from school had helped me see the flow of the argument. I could not wait for Ms. Soucey to hand back the paper so that I could read her comments. I wondered if she would notice the improvement I'd made.

When she finally returned the graded papers, I held my breath as I read through the comments and looked for the grade. My entire face lit up with sheer amazement when I saw the "A" marked at the top of the page, and I looked for my name on the title page just to make sure I'd picked up my paper and not someone else's. My face lit up with glee. By the time Mom arrived to pick me up after school, I was still unable to contain my excitement. I had barely shut the car door before I gave her the good news.

During my sophomore year, I took American Literature with American History. We were introduced to American writers from Hawthorne to Fitzgerald and Frost. We read plays like Arthur Miller's *The Crucible*. These stories transported me to different times and places. Using the stories and descriptions from the literature readings, I was better equipped to understand the abstract concepts discussed in history.

While American Literature helped me learn history, Art History complimented both history and literature, especially when we were studying World History as juniors.

Ms. Jueling, the Art History teacher, always provided historical and cultural context for the works of art we were studying. She often gave the history of a statue or building as a point of reference.

We used a historical timeline with pictures of the art so that we could see the connection between historical and intellectual developments and a particular movement in art. But we didn't just see the timeline and the art; Ms Jueling talked through all of the connections and asked us to put ideas and language of art into our own words. When a slide of Michelangelo's *Pieta* appeared on screen, she asked us to identify the artist, the date, the work, and then to identify what characteristics and techniques reflected the time period in which it was created. Her teaching style was a perfect fit with my learning style.

Ms. Soucey's teaching style, which used lecture and conversation about the material, also seemed to fit my learning style. However, even after the geometry meeting about my learning disability, there were still moments when I came across something that I couldn't do. When handed a test in World History, I picked up the test booklet and started answering the short answer questions and essays. I knew the material and had no trouble answering until the last page, a blank sheet with the instructions to draw a map of Africa, drawing in the countries and labeling them.

> " Even after the geometry meeting about my learning disability, there were still moments when I came across something spatial that I couldn't do. "

Shocked, I knew that no matter how hard I tried, the map problem might cause me to fail the test because I could not draw a table freehand, let alone a detailed map of Africa. I went up to her desk, my eyes beginning to fill with tears. Ms. Soucey saw the panicked look on my face and took me into the hall. I explained that I thought I couldn't do it. She told me to give it a try anyway and if I failed miserably, she wouldn't count that section. I was relieved that she didn't expect me to be a cartographer. I felt that the understanding she had gained about my spatial problems had helped her be sympathetic and supportive. She knew now I was capable and wasn't using my disability as an excuse.

About this time I realized that if I wanted to go to college, I would need to take SATs. I dreaded these tests because I had difficulty with answer sheets that had rows and rows of bubbles to fill in. I had a good chance of filling in the wrong bubbles even if I knew the correct answer.

Fortunately, because of my disability, I was granted extended time, and I could use a ruler, a calculator, and a blank sheet of paper. I managed to get through this grueling test with acceptable scores. As expected, my verbal score was much higher than my math score, but that didn't matter.

Although French proved more difficult than either History or English, I continued hoping that I would move past the grammar and focus entirely on literature. Even though I generally knew the meaning of the verbs, nouns, or adverbs and the proper pronunciation, I still had trouble placing the correct ending on a verb or placing the adverbs or pronouns in the right part of a sentence. This required attention to visual detail and was very difficult.

Gradually, my French class changed, and the conversation and instruction were in French. To my surprise, I found that hearing the lessons and class discussion in French allowed me to sort through some of the grammar.

At the end of my junior year and during my senior year, we began to move away from the grammar and started to read literature and write papers. Glad to get away from the drills and memorizing spelling rules, I jumped at the change. Besides,

> **" Once the props were set and the staging arranged, we made our entrances and began with the narration. "**

this was right up my alley since I had no trouble with reading or pronunciation.

One of our first assignments was to adapt several chapters from *Le Petit Prince* into a script and perform them in front of the class. We were divided into small groups, and each group acted out one chapter from the book. In our group, we began by translating the text into English and clarifying any words or verb tenses that were unclear. Then came the big day, the debut of our show. All of us had agreed to be responsible for our own costumes, and we came to class in character. Once the props were set and the staging arranged, we made our entrances and began with the narration. Laughs from the audience erupted and put us at ease. The rest of the skit went well with everyone having a good time.

As we read aloud *The Bald Soprano* in class, our teacher, Madame Ottom, clarified any vocabulary and idioms that created confusion. From the moment we started the play, the entire class was laughing out loud at the appropriately ridiculous scenes and characters. We liked the play so much that we decided to put it on for the entire school during the International Day talent show. Madame Ottom encouraged us to produce a translated version, but the class voted against changing anything and thought the play would lose its comic qualities. The curtain went up and we acted out the funniest scene we could find. We barely made it through without laughing while the audience members were quiet with puzzled expressions on their faces.

Senior year, in French IV, we were introduced to a multitude of French plays and class became almost like English class. Madame Ottom introduced us to the "Theater of the Absurd" and French existentialism. We wrote essays on the plays that Madame Ottom assigned, and I realized that I

Britt: Moving Forward, continued

had to apply my strategy for writing English essays to French essays. I thought about the organization just as I would for any other paper; however, writing the paper would be another issue. I wrote the paper, trying to pay close attention to the grammar, verb tense, and spelling.

The paper came back a week later with two different grades printed boldly in red ink at the top. For content, I received an "A" and for grammar a "D." Well, I thought to myself, I may not be the best at grammar, but I certainly understood the subject matter!

I went straight to Madame Ottom and asked if she would help me go over the paper and discuss aloud some of the mistakes. This process helped me and allowed me to develop a sense of how the language should sound rather than just relying on written distinctions.

Although eligible to take Advanced Placement English, I decided to take an elective on American romantic writers. This class appealed to me, and I had absolutely no intention of placing out of Freshman English at college. We read Thoreau, Whitman, Dickinson, Emerson, and Poe. Inspired by Walt Whitman's "Leaves of Grass," the class begged the teacher to go outside for a nature walk. The first scheduled day was postponed due to rain and rescheduled. On the second try, a heavy January rain again poured down from a slate gray sky. Fifth period arrived with no break from the rain and the low clouds wrapped themselves around the treetops. Still we sat in class eagerly anticipating our trek outside.

"All right everyone, it looks like we won't be able to take our nature walk after all." The teacher pointed toward the window where the drops of rain had become so numerous that they merged into heavy streams of water that ran down the side of the windowpane. Gusts of wind occasionally blew the rivulets of rain into swirling patterns. Sighs and murmurs emanated from the desks. Then at the same time everyone said, "Let's go today!"

"In the rain? We'll be soaked!"

"That's okay! Whitman and Thoreau would love seeing nature in the rain. Besides, it's the last period of the day, and we can all dry off when we get back," a student replied.

"All right, get your coats! You have five minutes. We will meet at the entrance to the basement tunnel," the teacher informed everyone looking at his watch. A loud cheer and a sudden rush for the door followed.

Rain beat on our faces and trickled down our cheeks. Joy and enthusiasm for nature and life filled us to the very tips of our toes, keeping us warm. We raced down the sidewalks and jumped in as many mud puddles as we could find. We pointed out birds and worms, generally taking in nature on the small and large scale.

> *Joy and enthusiasm for nature and life filled us to the very tips of our toes, keeping us warm.*

Impromptu quotations from Whitman and Thoreau sprang forth from the flock of gregarious students now traipsing through the downpour. We suddenly knew on an intimate level what Whitman meant when he wrote his words so long ago.

The hour ended too quickly, and we headed back to school with our clothes soaked and our hair wet to the scalp. Passing the wide-eyed stares and gaping mouths of teachers and students in the hallways, we encountered plenty of laughs about our sorry state. We came away with an experience none of us will ever forget.

Britt: Working

During high school, I decided to get a summer job. I had started to outgrow the summer school classes I had taken in middle school. Because academics during the school year were strenuous, I began to look forward to the break from the demands of intellectual life. My stepdad suggested that I apply to the television/radio station where he worked. He wanted me to get a behind-the-scenes look at how a radio station operated on a daily basis. Since I had an interest in broadcasting, I took his suggestion to heart and wrote a note to the president of the station, stating my desire to work for the summer and the date school ended. My letter was passed on to the Human Resources Department. Within two days of writing, I received a call asking me to come in for training the day after I finished school.

"We're going to train you in all of the departments—news, radio, sales, promotion, mail room, human resources—really anywhere we need help," the human resources director said.

I had been on many tours of the station before, and I knew my way around the building. But I still wondered if I'd be asked to do something I couldn't handle.

My first assignment was to train on the switchboard. I stared at the bank of blinking buttons and thought that I'd never be able to keep all of the lines straight. Thankfully, one of the receptionists took on the job of training me. I watched and asked questions and caught on quickly. Basically, each blinking light represented a call on hold. After I answered the call, I pressed the transfer button and dialed the appropriate extension. A list of the extensions for all the main departments kept us from confusing the different extensions.

> **" I stared at the bank of blinking buttons and thought that I'd never be able to keep all of the lines straight. "**

We were also equipped with a directory to look up individual extensions. I'm sure that I might have misdialed an extension or two the first day, but in general I managed to direct the calls efficiently.

We also answered general questions about radio and television programming and gave out information on the various community events and public affairs programs that the station frequently sponsored. Above all, we were polite at all times. Switchboard shifts were divided up into three parts. The early shift was from 7 a.m. to 3 p.m., the second shift from 9 a.m. to 5 p.m., and the third shift from 3 p.m. to 9 p.m. The person working the first shift opened the switchboard and unlocked the lobby, while the person closing made sure that the switchboard transferred over to the night service. We greeted everyone who passed through the lobby and directed them to the different departments. Security was another part of the job. We had different procedures for situations like airwave takeovers, protesters, fires, and bomb threats. At first, I thought that these were regulations that all television and radio stations were required to post. I never thought that I might actually need to use them. Then I began to hear stories of bomb threats against the station and personal threats against reporters and anchors.

After I started college, I came home for winter break and signed on to work a few hours a week. On January 16, 1991, I watched from home as the station broadcast the beginning of the Gulf War. The next morning, I arrived at work and learned that the station had received so many phone calls the day before that the lines had jammed and caused the entire system to break down. By the time I arrived at work, the system was up and running again. I jumped in and started answering calls. Most people

Britt: Working, continued

wanted to express their opinion about the conflict, some wanted to know why their favorite soap operas had been pre-empted, and some wanted to know if their relatives in Israel were safe. By the end of the day, I had answered over 500 calls. Just as I was about to leave, I found out that a group of anti-war protesters were marching toward the station, and I was instructed to lock the main doors. Just as I had secured the locks, my stepdad marched through the lobby with a camera crew right behind him and demanded that I open the doors.

"Open the doors!" he demanded after realizing the building had been secured.

"Not on your life! I have orders to keep the doors locked," I responded.

"I'm just going to take the camera crew out to interview the protesters. All they really want is a chance to express their view on-air," he replied.

I decided that reality took precedence over regulation. I opened the doors.

In May after I had finished my first year at college, I went back to work at the station. In July when the human resources coordinator told me that I'd be going to the Accounting Department, I thought she might be joking. "They want

> **" I just wanted to thank you for the wonderful job you've done in the Accounting Department. "**

me to work in the Accounting Department?" I thought to myself. This situation was a disaster just waiting to happen. When I walked into the Accounting Office on the second floor, I let out a sigh of relief when I found out that they really just needed help straightening files, sorting invoices, and stuffing envelopes. Although not the fastest filer, I knew that I could manage. That first day, I dove into work and realized that the filing system was a little more complicated than just alphabetical order. When I had questions, I asked and had them resolved without a problem. What I did not expect from the Accounting Department were parties. Almost every day they found an excuse to have cake, ice cream, or cinnamon rolls.

I could not believe that I had survived three weeks in the Accounting Department. I only had another week before I was scheduled to head back to school when the head of the Accounting Department called me into his office. I thought that I'd made a horrific error costing the station money. I tried to think of what I might have done wrong: reimbursement checks with too many zeros, numbers transposed, important information misfiled. I opened the door slowly and took the seat across from Mr. Lewis and waited for the bad news.

"Britt," Mr. Lewis began.

"Oh, no, here it comes," I thought to myself.

"I know you're going to leave us to go back to school, and I just wanted to thank you for the wonderful job you've done in the Accounting Department."

Shocked, I struggled to say something, but before I could say anything Mr. Lewis continued, and in all seriousness said, "I think you should consider a major in accounting."

ಙ ಙ ಙ

Julie: Preparing for College

While teachers in middle school thought I was setting my sights too high for Britt, many of her high school teachers were encouraging about Britt's academic prospects. Math and other kinds of spatial tasks continued to be difficult, but she excelled in courses that were language-based. Britt's teachers saw her as an intellectually capable young woman, and they encouraged her to look at selective colleges in other parts of the country. "It doesn't hurt to set high goals," her college counselor said, "but be sure to have a backup, just in case you don't get into one of your top choices." No, she wouldn't be going to MIT, but a small liberal arts school would be a good fit for Britt. Such a school would help her develop her strengths in a setting with more support than most large universities provide.

During her junior year, Britt sent off postcards requesting application material from a number of schools. Before long, mountains of catalogs and brochures began to arrive; soon one whole corner of Britt's room was stacked with materials from colleges.

One rainy Sunday evening in April, when we were gathered at the dinner table, my husband asked Britt if she had received any college information during the week. As I went to the kitchen to get the dessert, Britt retrieved the college mail from the past week. She returned with a stack about ten inches tall and placed it on the table. Britt picked up the catalog on the top of the pile. "This place is entirely too proud of its football team," Britt said leafing through the view book. "I hate football."

"Most schools have football teams," I said. "You don't have to watch if you don't want to."

> **" Before long, mountains of catalogs and brochures began to arrive. "**

"But not all schools devote six pages in the view book to the football team," Britt replied

I had to admit she had a point. Britt picked up a course catalog from a liberal arts college. "Look at all the history courses they have," she said.

Britt read through five pages of history offerings. "I can't believe all the history courses. I want to take all of them."

"Well," John said, "Where do you want to go to college?" My husband was delighted with Britt's enthusiasm about the college view books.

"I don't have any idea," she replied. "Where do you think I should go?"

"I think you should go to school on the East Coast—no, New England. How about Dartmouth?"

"You went there, right?" Britt asked. "Then I know it's not for me." Britt made a face and started to get up from the table, indicating that she did not intend to have this argument.

"Britt, I'm trying to be helpful. Dartmouth is a good school," John said in his usual take-charge way. "Okay, if not Dartmouth, how about Smith? You should like the fact that Smith women wouldn't go out with me."

"John," I interrupted, saying his name firmly, "there are good liberal arts schools in the Northwest that aren't quite so hard to get into." I was afraid John was setting Britt up for disappointment.

"It doesn't hurt to look," John said. "Let's plan a trip to visit East Coast colleges over the summer. Britt, you come up with a list of schools that look interesting, and then we'll talk over the list and plan a trip."

Over the next month, I helped Britt make a list of the criteria she wanted to look for in a college. I also found a book in the

The Source for Visual-Spatial Disorders

Julie: Preparing for College, continued

library that listed colleges with programs for learning disabled students. Britt knew she didn't want to go to a school that would single her out, but she also knew that she would need some accommodations; she needed a school that would offer some kind of support.

But there were other issues I worried about. How would she deal with being far from home and her support systems? How would she manage the airports, buses, and trains that might be necessary to get to a school in an isolated area? How would she find her way around in an unfamiliar town?

One evening after Britt had compiled her list, we spread the view books out on the dining room table. John was convinced that Britt should go to the best school she could get into and for him that meant Ivy League. I looked on while Britt and John shuffled through the college materials.

"Okay, Britt, how about going to New York and Boston next summer? We'll rent a car and drive up to Hanover. You need to at least look at Dartmouth."

Britt made a face. "There is no way I'll go to Dartmouth. You've talked too much about your disgusting fraternity and the parties that lasted for days." John extolled the merits of Dartmouth while I was worried about Britt going to college far from home. I kept my worries to myself. There was no sense putting up barriers for Britt before she had a chance to explore her choices. Soon John was making plans for an East Coast college tour in July.

Britt was excited about looking at the colleges. I decided to put

> **"** *No matter where she went to school, at some time in her life Britt would have to negotiate airports.* **"**

away my anxieties about disappointments. Instead I began to see the college trips as a way to help Britt become an independent traveler. No matter where she went to school, at some time in her life Britt would have to negotiate airports. To do that, she needed to put the "airport" into language so that she would not see it as an impossible maze.

The late July college trip did not begin well. None of us were ready to get up when the alarm went off at 4:30 a.m. to take the morning non-stop flight to the East Coast. Grumpy, we arrived at Sea-Tac airport by 6:00 a.m. for our 6:45 flight to New York. As I stood in the line to check luggage, I started pointing out a variety of features, putting the airport into language. "This is the United check-in area, but if we kept going we would get to Alaska, Northwest, American, and Contintental." Although we didn't use it that day, I explained "Curbside check-in." I pointed out the signs that said "Gates" and the departure and arrival screens. I explained the numbering system for gates, the escalator to baggage claim, and the underground train that would take us to the North Satellite, noting that though airports were different, they were laid out in much the same way. I modeled how to ask directions from airport personnel and accompanied the modeling with words, always with words. "Britt, you can always ask one of the people at a counter for directions," I said. "Let me show you."

I knew that if she were tired or stressed she could "feel" as if she were lost, and that feeling would lead to panic. As we went down the escalator, through security checks, and into the corridors that led to the gates, I was talking to Britt about where we were and where we were going, always tying images when I could to what I was saying. Once we were in the North Satellite, I continued talking as we read the "Departing Flights" screen and made our way to the gate. Inside the plane, I talked out loud as I looked for our seats.

Julie: Preparing for College, continued

Our plane landed at JFK, and we poured into the already crowded concourse. I asked Britt to make sense of the airport signs, asking her to lead us to baggage claim and then on to ground transportation.

Moving through the crowds, Britt spoke the directions as we followed her lead. "There's the sign for baggage claim," she said. "Oh, there's another one. I guess we're going in the right direction." With some help, Britt led us to baggage claim. While John stood in line at the rental car counter, I showed Britt "Ground Transportation" and explained to her how to read the signs on the busses and airporters. Although we had a rental car for this trip, sometime Britt would need to know how to negotiate these other forms of transportation, and she would need to put this type of space into language.

By the next morning, we were in Massachusetts for our first campus visit and tour. As we walked around the campuses with Britt, I tried to keep up the running commentary that would put the spatial elements into language. I knew that once Britt had the map created by language, she would remember it much better than I would. After the college tour, Britt looked pensive as we walked back to the car.

"Well, how did you like it?" I asked.

"It's okay," Britt replied. "I just can't see myself going to school here. I don't know why. I just can't."

Later that afternoon we arrived in Northampton, Massachusetts, the home of Smith College. I put the town and the college into language she could use. I tried to do it conversationally. "Oh, Britt this must be the main road to town—it's four lanes wide and it looks like a bus route. At least I see a bus coming up the hill. That's a really attractive brick building across the street; it must be a dorm. Yes, there's a sign that says Northrop Hall."

Britt joined in, "Look, there's Saint John's across the street. It's an Episcopal Church, so that's probably the church I'd go to if I came here." Sentence by sentence, Britt began to build the map that would allow her to become independent, even when she was thousands of miles from home.

When Britt returned from the tour of campus, she was quiet as she walked to the car.

"Well, how did you like it?" I asked as I closed the car door.

"I didn't like it," Britt said. "I

> *Sentence by sentence, Britt began to build the map that would allow her to become independent, even thousands of miles from home.*

loved it. This is where I want to go to college."

"What did you love about it?" John asked.

"The campus, the buildings, the dorms, the girl who showed me around. Everything." Britt continued to talk about the classes as she waved a course catalog.

I hoped she wasn't in for a big disappointment. Smith was and still is highly selective. Could Britt get in? If she did, would she be successful? How would she manage in such a competitive environment so far from her support systems?

જી જી જી

Britt: Finishing High School

By the fall of my senior year, I was driving myself to school one day a week. Mom decided that she would work from home on Fridays, and I was on my own for the hour commute to and from school.

Although having a car at school should have been liberating, it became an added stress. Exhausted after an entire week at school and full day of classes, I dreaded the drive back home. Tired, I struggled to concentrate on the road, constantly talking my way through the drive to keep myself from making any hazardous mistakes. I recited landmarks on my way to the freeway—Stadium High School, the old Elks Club near the turn onto the freeway, and finally the signs for the freeway itself.

> **" I remained a little apprehensive about going so far away from home and wanted to make sure I was making the right decision. "**

Merges were tricky, and I found that I talked myself though the points in the merge when I needed to accelerate and decelerate according to the flow of traffic. I always heaved a great sigh of relief when I finally settled into a lane and did not have to worry about judging the speed of other cars as well as judge whether or not I had enough room to move into another lane. I made my way home through the traffic on the freeway, the streets of Seattle, and finally home. When I arrived home at last, I collapsed on my bed for a well-deserved nap before dinner.

In the late fall, I found myself telling Mom and John that I just could not manage the commute any longer; it was becoming too hard and draining. Since we had already discussed the possibility of boarding part of senior year, I pled my case. I wanted to have the opportunity to participate in the spring musical "Canterbury Tales" where there would be numerous rehearsals, and with graduation approaching, there would be evening and even some weekend events I wanted to attend. Being at school would make everything easier. I could come home some weekends and Aunt Jill was nearby if I needed anything. Plus, I would be leaving for college next fall, and this would give me a chance to experience life away from home. The decision was made. I would board the last half of my senior year.

There was never any doubt that I'd go to college. I applied to several small liberal arts colleges on the West coast as well as Smith College in Massachusetts. Although Smith was my first choice, I remained a little apprehensive about going so far away from home and wanted to make sure I was making the right decision. Still I went about the final half of my senior year hoping that I would get into Smith.

Every spring the Upper School had a free day called Mountain Day. Students are advised a month before Mountain Day to start bringing casual clothes to school. No one except the student body president and headmaster knew when Mountain Day would occur until the day it was announced.

One April morning, I made my way to the auditorium for a mandatory assembly on fire safety. I hoped that we would return to classes quickly because I had a paper for American Romantic Literature due that day, and I needed to use my scheduled study hall so that I could turn in the paper on time. As we entered the auditorium, I sat down quietly. Then the deep red curtains on the stage began to gently sway and pull apart revealing a gigantic banner stretched across the entire length of the stage that read "Happy Mountain Day!" An explosion of excited cheers and shouts emanated from the seats. This was definitely going to be a great day. The

The Source for Visual-Spatial Disorders

Britt: Finishing High School, continued

initial thrill subsided and I began to feel a knot forming at the pit of my stomach. Oh, no I thought to myself. What am I going to do about the paper? It's almost done and now I have no chance of finishing today. I knew that I'd have to turn in the paper, Mountain Day or not. I began to plan out my argument for a small extension, preferably the next morning.

Once everyone began to quiet down, the student body president announced where and how we would spend our free day. We had thirty minutes to change and gather everything we wanted to take with us. Buses were waiting outside to take us to the lake where a barbecue lunch would be provided. The dazzling warm April weather signaled the start of a spectacular day.

I ran upstairs to my room to throw on some jeans and a T-shirt and grab some sunscreen as well as a pair of shorts and a sweater. Throwing these into a bag, I ran down the main stairs and began searching for the literature teacher.

Still worried about the paper and about ruining my grade in that class—I still needed to maintain a decent grade point average if I wanted to attend college—I found my professor in the front hall in conversation with other students who were also worried about their papers. We pled our cases and managed to have the deadline extended until the next morning. With a sigh of relief, I felt free to enjoy the rest of the day.

We arrived at the park and set up the picnic site with snacks and sodas. Everyone spread out and found different activities to take part in. Some went down to the lake to swim, some walked around the small hiking trail nearby, some people found

> **" I walked around the hiking path and thought about next year. "**

a place to play tennis, some started a game of beach volleyball, and others played Frisbee on the lawn.

I walked around the hiking path and thought about next year. I was anxious to hear from schools. I had carefully studied all of the colleges that I had visited my junior year and had pored through countless brochures. Knowing that I had been accepted at two colleges made me feel confident and a little more relaxed about the last semester at Annie Wright. I thought about what my life would be like in college and whether I'd decide to join a sorority. I also could not help but think about May Day and graduation that was less then a month away.

My concentration was broken when I arrived back at the picnic site greeted by the wafting aroma of barbecue cooking. Mr. Timson, the biology instructor, sat slowly and methodically turning hot dogs and flipping hamburgers and doling them out to eager hands. I stood in line and marveled in the moment.

After lunch we packed up our gear and prepared to return to school, tired but happy from our day in the warm Northwest sunshine. Everyone was in a good mood. The end of the school year was in sight, and the nice summer-like weather reinforced that fact. We piled back onto the bus and returned to school around 3 o'clock, just in time to hear the last bell ring.

Feeling fortunate after the day off, I returned to my room to find a little plant at the base of my door. "Hmm," I thought, "what is this all about?" As I took a closer look, I noticed a phone message from a pink While You Were Out note pad placed next to the flowers. It read "Your mom called at 11:57 a.m. on April 4th. Message: You Got into Smith!" I was so shocked I didn't know where to go or what to do. I ran down the hallway as fast as I could to the phone booth. I called Mom and as soon as she answered the phone I peppered her with questions. I was too excited

The Source for Visual-Spatial Disorders

Britt: Finishing High School, continued

about the letter to even think straight. As I said goodbye, we agreed to meet for dinner and a celebration. Unable to contain my exhilaration and thrill at finally being accepted to Smith, I rushed downstairs where I hoped to find my Aunt Jill and Ms. Soucey, my advisor, to share the good news.

"Guess what? Guess what?" I yelled as I burst into my aunt's classroom. "I got into Smith!" I said, trying to keep from shaking with excitement.

When my aunt realized what I was telling her, she gave me a great big hug. She knew how hard I had worked through high school and was just as excited as I was about the news.

April turned into May and with all of the college decisions behind me, I began to experience senioritis. A week away, waiting for May Day and graduation seemed like an eternity. All I had to do was get through my exams and write my last paper, and I was free.

Before I knew it, classes were over, exams had been graded, and graduation weekend had arrived. The weekend was filled with events like the senior luncheon, a chapel service, and finally May Day and Graduation. The night before May Day, I chose to stay at school. Too excited to sleep, the seniors and juniors stayed up most of the night talking and watching movies. As tradition stated, the juniors woke up the seniors May Day morning, banging pots and pans, blowing whistles, and knocking on doors, telling the seniors to get up. The juniors had prepared breakfast for everyone and had started to decorate the lawn with flowers for the ceremony.

May Day, a long-standing tradition at Annie Wright, was a way to celebrate the seniors and alumnae. As seniors, we chose a May Queen, the student we felt best represented the class. With a great deal of pomp and circumstance the seniors were presented to the audience. The May Queen made a speech after which students from lower and middle school presented a pageant consisting of musical numbers and dances.

Graduation followed soon after. On Graduation Day we lined up in the senior hallway, waiting to go into the chapel. Ms. Soucey came down the line with bobby pins, making sure our caps were attached to our heads. Clad in graduation robes, the seniors marched in alphabetical order to the chapel filled with streams of colored lights that gave off a soft glow. Several seniors, including myself, had been preselected to give commencement addresses. As the first two speeches were delivered, I began to feel tense. My speech was next. I had chosen to center my speech on Walt Whitman's "Open Road." When my turn arrived, I went up to the podium and with a quavering voice, I started to deliver my address. With the first few lines out of the way, my nervousness subsided, and I began to relax, focusing on the message I was trying to convey instead of my anxiety.

Once the speech was over, I was free to enjoy the rest of the ceremony. At the end, Mom, John, brother John, Sarah, and Carol were all there to congratulate me. I headed out onto the front lawn of the school and with the other 30 graduates had our picture taken. Then with a loud triumphant cheer, we all threw our graduation caps into the air.

ಐ ಐ ಐ

Carol: Why Did Hard Things Get Easy?

On the way home from Britt's graduation, I had stopped to pick up some papers this balmy Sunday afternoon and found Margaret in her office finishing a report for a Monday conference. Britt's graduation was such an emotional high that I was having difficulty coming down to earth after the beautiful services in the chapel at her school. I wanted to share my joy with Margaret, who had also taught Britt at the clinic—to tell her how competent Britt looked, marching up to get her awards.

As I stuck my head into her office, Margaret waved me in. Soon we were reminiscing about Britt over steaming cups of coffee. We shared many memories of her determination to overcome the mysteries of mathematics. As I pushed back my chair to leave, Margaret said, "There's something I want to ask." She paused and then said, "Isn't it wonderful how school got so easy for Britt? Why did things that were so terribly hard get easy?"

Even after taking extra breaths, my voice betrayed irritation, "Who said they are easy?" I challenged Margaret who had ventured that word. "Who said that? Easy is—is—Easy!" I said, my voice rising. "Easy is—is falling off a log." Letting the cliché express my annoyance, I sat back down and added, "Britt has to work like any other student, in fact harder than a lot of them because she has to do things her way, and that is *not* easy."

"Carol, take it easy," Margaret smiled at the accidental word play, "I only meant she flat out

> *" Britt has to work like any other student, in fact harder than a lot of them because she has to do things her way, and that is* not *easy. "*

couldn't do math and now she can." Her words tumbled out in response to my irritation. "Britt couldn't organize a paper or take tests." Margaret sat forward in her chair and drummed her fingers on the table between us. "You know what I mean. She's driving, for goodness sake. And she's not getting lost. What happened?"

My reply was deliberate, "I suppose that you are really doing me a favor by asking these questions," pausing and then adding, "I have been trying to put all of the pieces together. Only there are so many factors—even now I feel like I have to qualify everything that I say." I added, "Actually your questions might help me cut through philosophizing to specifics."

Silence settled over the room. Margaret sipped her coffee and finally said, "Well?"

My words tumbled out, "Britt created a 'frame-of-reference' for each spatial task. She put that reference along with her actions into her own words. Then she could use her words to guide herself when she had to do those spatial tasks." I heard myself speaking faster and faster. "And she made images to recall the process for next time. Frame of reference, language, and images— she needed *all* three things, only sometimes they were in a different order."

Margaret looked dazed. She stared at a distant point over my left shoulder. Time ticked away before she sighed and said, "You're probably right, but I need more explanation."

"This whole thing is not easy to explain," I said. "Maybe if I tell you some stories," I proposed, "but stop me if the story doesn't make sense. *Spatial* means relationships and comparisons. Britt couldn't make comparisons that involved scale or describe things until she understood relationships like 'how big is an inch? How does that compare to a foot?'"

Margaret nodded.

I continued, "She could count and measure with a ruler, but

The Source for Visual-Spatial Disorders

Carol: Why Did Hard Things Get Easy?, continued

the number didn't mean much to her until we did that bit where she stuck her finger in ink that was an inch deep.

> **" Language is a powerful organizer for nearly everyone. It is Britt's lifeline... "**

When she used her stained finger to measure many objects *and* talked about the sizes as she did so, she created a 'frame-of-reference' for an inch. She *knew* 'inch' and could quickly build up from there to understand 'foot' and other measures. Now when she said 'a foot has 12 inches,' the words conveyed meaning. She had all three things in place: frame of reference, descriptive language, and image. She could then 'get into' a problem that included scale in various ways. She might image it like most of us do, or she might talk her way into it."

I added, "Of course that didn't work until she had her reference down pat."

"But isn't driving far more complicated?" Margaret asked.

"No question, driving was a monster to tame," I responded. "So many 'spatial' pieces demanded attention: the size of the car, the location of the sides of the road and the center stripe, the position of other cars, the distance to the corner, the angle and width of the lane as it swung around a corner, the speed of the car in relation to overtaking other cars or stopping distance and on and on. But Britt and her mother rehearsed every small bit. They stepped off lane widths and talked and described until Britt 'knew' that monster."

"By 'knew' you mean....?" said Margaret.

"That she could image each part of driving and she could talk her way through the actions as she did them," I responded. "Language is a powerful organizer for nearly everyone. It is Britt's lifeline, the conduit of certainty for the road, the car, and the left turn across traffic."

I added, "She is a thoughtful, careful driver who earned her license. I guess that's why I reacted so strongly to you describing hard things as 'easy.' They are deserved. *Deserved* is a much more accurate word."

"I didn't mean to insult her," Margaret said.

"I'm too sensitive about that, I guess," I said. "It's just that Britt was 'proactive' before anyone coined that word."

"She built her frame of reference for shape and perspective by tracing objects on the Learning Window. Once she saw and described the difference between the three-dimensional objects and her tracings of them, she 'knew' and could transfer that knowledge into practical applications," I said. "That's how she learned to understand how point-of-view shifts when you move or when objects move. That was a big step, but she caught on so fast."

"Did you have to teach her everything? What if you missed something important?" asked Margaret.

I answered, "Britt had already figured out a great many things. I guess you would call them adaptations."

Smiling, I recalled for Margaret how Britt built her own form of modifications into her play. She organized her brother and friends into a skit, helped children in a store decide which clothes best went together, and made witch's brew from puzzle pieces instead of putting them together. She took control and modeled or acted out instead of depending on her faulty visual memory.

"Margaret," I said, "Britt and her mother had already created a very important bridge over her spatial problems before I ever met them. Maybe it was an accident. I prefer to call it 'intuitive parenting,' but by any name, it was very important."

"What kind of bridge?" Margaret asked, her curiosity apparent in her voice.

The Source for Visual-Spatial Disorders

Carol: Why Did Hard Things Get Easy?, continued

"Remember," I added, "Britt understands what she reads. She can and does build images in her head as she reads and as she listens. Most kids with spatial problems cannot do this. That's why they do not get meaning even when they sound good as they read."

"So how did that happen?" Margaret asked, looking very perplexed.

Her question triggered a host of memories. How could I convey the core reasons for Britt's reading success? I paused before replying, "Julie, Britt's mother, loves literature and wanted to share that love with Britt. However, she had a major problem. As a single mother with a demanding job, her time was very limited. So during Britt's early school years, they read aloud to one another while doing other things and they talked about the stories they read. Britt might have read to her mother while Julie made dinner. They read Britt's schoolbooks, and they even read Julie's students' essays. They talked about characters; they talked about plot. Britt is a whiz with oral language—both talking and listening. And Julie could then be certain that Britt understood content as they went along. By asking Britt to predict what might happen next, Julie reinforced Britt's ability to build images for the print she read. I am convinced that this activity helped Britt enormously, and in the most effective way, long before anyone knew she had a special need."

Margaret looked thoughtful as she said, "What a gift! What a special mother-daughter gift! They were just sharing what they loved with each other."

"Well," Margaret continued, "It's not really wrong to say hard things became easy. Britt is a wonderful learner and a good student. Of course, she has taken a different road than most students. And a longer one. Carol, you are focusing on all the hills and valleys Britt and Julie have traveled. Maybe you should focus more on the scenery they have encountered along the way. What a deep and personal journey for both of them."

She stood, gathered the papers she had put on a chair as we talked, and said with a grin, "Now what they need to do is write a travelogue for the rest of us who have kids with special needs. Tell them that from me." Her words echoed as she went to her car to head home to her own special learner.

      ಐ   ಐ   ಐ

> **" Now what they need to do is write a travelogue for the rest of us who have kids with special needs. "**

The Source for Visual-Spatial Disorders

Britt: Going East

In August, I packed my suitcases and trunk and headed for college in Northampton, Massachusetts, thousands of miles away from home. With the flurry of activity that surrounded graduation and a summer job at a local television station, I didn't have time to think too much about leaving for school until I was faced with the reality of packing the black steamer trunk with the things I thought I'd need to survive a year of college. I finally realized I was leaving and Northampton felt like the end of the Earth. An odd mixture of excitement and apprehension came over me as I began to worry about college and how I would manage so far from home. Instead of worrying, I threw myself into packing.

> **"** *I was faced with the reality of packing the black steamer trunk with the things I thought I'd need to survive a year of college.* **"**

On August 27th, we loaded up the car and headed to the airport to catch an early flight to the east coast. The trip to Smith turned into a family affair. My stepdad said that we could carry more suitcases if everyone went to Northampton. And it would be an adventure. We stopped in Boston for a few days to do some shopping since there were still a few things I needed.

After a short stay in Boston, we all piled into the rental car and headed down the Mass Pike to Northampton. I watched out the car window as we whizzed through the New England countryside, through towns with strange names like Belchertown and Holyoke and Chickapee. We passed the industrial town of Springfield. As we left the outskirts of Springfield behind, the craggy sides of the Berkshire Mountains unfolded, and I saw for the first time the expanses of lush green birch and maple trees covering the hillsides.

We arrived in Northampton and wound through the town; restaurants, shops, pharmacies, and bookshops lined Main Street. Once we arrived at Northrop Hall, and my family began to unpack the car, I checked in with the head resident to find my room. The building was Neo-Georgian style and looked similar to Annie Wright. I immediately felt at home. To my surprise, I was even greeted by someone I knew. Kim, a girl who had graduated from Annie Wright the year before me, was helping out with orientation and steered me to the head resident's suite.

After meeting the head resident, I headed to my new room on the third floor. When I arrived at the top of the stairs, out of breath and sticky from the humidity, a tall, slender, dark-haired woman greeted me at the entrance to our room. Chonira and I had exchanged letters over the summer, and I knew that she was from Sri Lanka, a small country off the southern tip of India. Chonira had arrived a week earlier in order to get settled and attend a program for incoming international students.

"You must be Britt! I'm Chonira, but everyone calls me Choni—it's easier to say," she said, holding out her hand.

"Hi, nice to meet you," I replied and introduced my family who were bringing suitcases from the car.

Our corner room had two large windows. My window looked over Elm Street to John M. Green Hall, a stately brick building with white pillars and ivy-covered walls. A magnificent maple tree just to the left of the window blurred part of the view. Choni's window looked over a large expanse of lawn to the student center.

As I started to unpack, we discussed what we should do with the room and how we should arrange the furniture. With only a few adjustments, we had the L-shaped room just right. The room looked bare, even with double sets of furniture.

Britt: Going East, continued

We had two beds, desks, phones, bookshelves, and dressers. Later, we would add a few personal touches to the room—a rug, posters, and family pictures.

On the last trip up from the car with boxes, I met two other freshmen moving into their suite just a few doors down the hall. Mom and I greeted Louisa, a quiet, calm girl from Quebec, and her mom. Louisa invited me in to see her room and meet her suitemate, Bronnie, a lively, vivacious field hockey player from Pennsylvania who seemed to be the exact opposite of Louisa. We sat around and talked for a while, covering the basics of where we were from, our families, and our first impressions of Smith. I liked them right away and knew that these girls would become my friends.

After the suitcases, boxes, and trunk had been carried up to the room and unpacked, Choni took us on a tour of the house—a basement with washers and dryers, a storage area for unused boxes and suitcases, and a lounge with old vinyl sofas, large armchairs, and a long study table. The first floor contained a television room, living room with a piano and a fireplace, and the dining room.

Later that afternoon, we went to the opening convocation. The entire freshman class crowded into John M. Green Hall and the late August humidity weighed heavy in the air as people fanned themselves with the programs in a vain attempt to cool off. Despite the heat, there was a palpable sense of excitement and anticipation. I looked around the grand auditorium and knew that the walls were full of history and wondered how many women had sat in that very room.

The College President, Mary Maples Dunn, energized us with a welcome speech. She told us that although we had all been carefully chosen to be at Smith, not everyone would make it to graduation. The academics were tough, and if we looked to our left and looked to our right, one of those students would be gone by the end of the first year.

I said goodbye to my family and spent my first night in my new room. Tossing and turning, I found it difficult to fall asleep. The breeze that came through the window by the bed barely broke through the heat that had built up throughout the day. I thought ahead and wondered what my new life in college was going to be like. My mind kept racing, and I started to wonder for the first time if I could handle being so far away from home and whether I'd fit in.

During that first week, Choni and I discovered that we were compatible. We both went to sleep relatively early, were quiet, neat, and liked to get up early. I was relieved to know

> "*Despite the heat, there was a palpable sense of excitement and anticipation.*"

that we would get along. We had a lot in common, and we became friends.

Over the next few days, we went through freshman orientation. Unfamiliar with the campus, I found that having a roommate, especially one who knew her way around, had advantages.

The Smith campus is one of the most beautiful campuses on the East Coast, with winding, tree-lined footpaths, a botanical garden, and a pond with a boathouse perched at the edge. As I walked among the academic buildings, I marveled at the pristine, picture-perfect setting.

During that first week, I set out to explore the campus and visit as many buildings as possible as a way to orient myself. I walked through the art museum; Seelye, College, and Wright Halls; and finally Nielson Library. I walked through the stacks in Nielson, selected a few interesting

The Source for Visual-Spatial Disorders

Britt: Going East, continued

books, and made my way down to the reading room where there were rows of long mahogany tables lined with lamps. The walls were covered from floor to ceiling with every kind of magazine, journal, and newspaper imaginable. I liked this building.

One of the assignments during orientation was to meet with our advisor and pick our first semester classes. Before arriving at Smith, all freshmen were assigned a pre-major advisor based on their academic interests. Smith had no core curriculum requirements, so our advisor was supposed to be our guide through the confusing maze of courses. My advisor was from the English department. To prepare for my meeting with him, I carried the thick course catalog with me everywhere, searching through the hundreds of available classes. I wanted to take them all with the exception of calculus and chemistry. How would I ever narrow them down to just four? I hoped my advisor would steer me in the right direction.

I arrived at my advisor's office in Neilson Library and waited out in the hallway for my turn. I noticed the copies of my advisor's articles from the *New York Times Book Review* section and knew that I was in the big leagues. I grew increasingly nervous as I waited. This was the first step in my college career. I was deciding my first set of classes and felt this would decide the rest of my academic career right there in the basement of Neilson Library.

When I got into the office, my advisor asked me a few basic questions and then asked me which classes I was interested in taking and what I thought my major would be.

"I'm having a little trouble deciding. I know that I want to stay away from calculus and science," I said, explaining that I had a learning disability and wanted to get my bearings before attempting anything that might be too difficult without support.

"Well, we are supposed to encourage students to try different classes. You could take a sampling of classes from the humanities and arts if you like," the advisor offered. He sounded as if he were offended that I wasn't going to be an English major. "Think about it some more, and once you've made your decision, I'll sign your registration card." This was his final piece of advice, and I was dismissed to make the decision on my own.

After a great deal of thought, I finally settled on a suitable schedule—French II, Art History Survey, English: Fact or Fiction, and American History. I thought that this would be a good mix and help me try a broad range of subjects and departments.

Art History Survey, or Art 100 as it's known around the campus, turned out to be one of my most enjoyable classes freshman year. Art 100 was not just a class, it was a production that took on a life of its own, seeping into every aspect of a student's life. Almost everyone at Smith took Art 100 or at least audited the course. There were even T-shirts that said "I Survived Art 100." The class had around 300 students and relied almost entirely on lecture. At dinner, I heard desperate Art 100 students asking for class notes or to be quizzed.

Pictures of the slides were taken from each lecture and hung in a special room in the art library commonly known as "The Wall." The term "I'm going to the wall" usually sparked at least a sympathetic

> " *I noticed the copies of my advisor's articles from the* New York Times Book Review *section and knew that I was in the big leagues.* "

nod. Without a book to study from, frazzled, bleary-eyed students spent hours staring at "the wall," attempting to memorize the name, artist, date, artistic movement, and place of origin just in case it might show up on the exam.

Art 100 would have been my hardest course, but thanks to my high school art history teacher's style, I had a head start and found the course to be my easiest. In fact, Art 100 sections turned out to be the best part of art history. There were only 20 to 25 students in each section. Mr. Felton, one of the most animated faculty members in the Art History Department, taught our section and encouraged us to voice our opinions about the artwork. He even made ancient Greek and Roman art come alive. In an attempt to demonstrate the movement in a Greek statue, Mr. Felton stood in front of us and with a flourish mimicked the statue's pose perfectly.

We had a number of English classes to choose from. "Fact or Fiction" jumped out at me basically because of its intriguing title. I liked the reading list. We were assigned a wide range of materials from poetry to novels and even a play.

I did not realize I would have difficulty with paper writing. As a high school student, I had been proud of my improvements in writing and even looked forward to paper assignments. But college was a little different. I found myself struggling for time to finish all of

> **" *I did not realize I would have difficulty with paper writing.* "**

the reading and write the papers. Since I had not learned the value of writing my papers directly onto a computer, I still had to write the paper in long hand and then type the paper, adding an incredible amount of time onto the entire process. I barely had time to edit my papers by the time they were due. I knew that I could improve my papers, and I tried to get appointments in the writing center. With only a few writing advisors available, getting in to see one was almost impossible.

As the semester moved forward and I began to get my papers back from the professor, I noticed that there were lengthy comments lining the margins of the paper. The teacher went into great detail about how I needed to improve my grammar and spelling and that it was getting in the way of the material I presented in my papers.

At this point, I began to wonder whether I would make it at Smith. The weather had turned cold and dreary, and I faced my first set of finals. I was convinced that everyone was getting better grades than I was, and I felt that my attempts at higher academics had been a total failure. I knew that Choni, Louisa, and Bronnie were happy with their classes and seemed to be well-liked by their professors. I was used to having a supportive faculty, and I felt very much alone.

I was too busy the first few months to feel homesick. But once I was back home at Thanksgiving, I did not want to go back to Smith. I could hardly bear to get on the plane back to Massachusetts, and I cried through the entire six-hour plane trip.

ಙ ಙ ಙ

Julie: Parenting a College Student

By late August, Britt was packed, and we were on our way to New England. We arrived in Northampton a few days before the residence houses opened. We needed to buy the things that we didn't want to ship across country and we wanted to have time to help Britt become familiar with her new surroundings. We talked our way around the campus itself and around Northampton. We helped Britt find a bank and open a local checking account, and we identified a pharmacy, grocery store, variety store, and place for haircuts. My goal was to leave Britt with a language map of the town and the campus so she would not get lost.

As John and I drove out of Northampton, I felt confident that Britt was in a good place—she was making friends and enjoying the orientation activities. But by the time the plane was in the air, I was sobbing, full of self-doubt. Maybe a place closer to home would have been better? What if she fails? My husband, seeing my obvious distress, suggested I use the airphone to check on Britt. Although Britt wasn't in, we left a message and somehow I felt better. Evidently, she was doing better than I was.

During the first month of school, we talked to Britt about once a week, and she seemed to be doing fine. She had made good friends in the dorm and seemed happy with college. By October, we started to hear from Britt more often and the calls were often tearful. One afternoon, I was in my office intent on finishing an agenda for a 4 p.m. meeting. It had been a difficult day with class,

> **" By October, we started to hear from Britt more often and the calls were often tearful. "**

student appointments, and a budget proposal to complete. My desk was six inches deep in mail and unread student papers. Annoyed at yet another interruption, I picked up the phone on the first ring and answered in my "I'm really, really busy" voice.

"Hi, Mom," Britt said.

"What's up?" I asked, trying to discover the reason for the midday call to my office so I could handle it and get back to the agenda.

"Oh, Mom, I just wanted to hear your voice."

"Are you okay?" I asked. Worry about finishing the agenda had turned to worry about Britt.

"Yes," followed by a sigh. "I have a lot of work to do before tomorrow." A longer pause.

"Do you want to talk about it?" I asked, my worry returning to annoyance. I glanced at my watch—3:50. The light on my phone began to blink, signalling another incoming call. My secretary was knocking on my door. I covered the mouthpiece and called, "Just a minute."

"No, I better let you go; you're probably busy."

"Okay, well, let me call you later when I get home." Another pause followed.

"Okay, but I'll probably be at the library," Britt paused and sighed. "Bye."

Our conversation left me frustrated, worried, and upset. I simply didn't know what to do. After a conversation like this, how could I focus on my agenda or the subsequent meeting or anything else for that matter?

That night when I called Britt I was met with the usual cheery voice: "Hi, this is Britt. Leave a message and I'll call you back." I left a supportive message, urging Britt to call me if she got back to her room before midnight. That night I lay awake full of anxiety, full of misgiving and doubt about Britt being so far from home without a support system that she could count on.

Eventually I slept, but I awoke with the knot in my stomach

The Source for Visual-Spatial Disorders

Julie: Parenting a College Student, continued

that is associated with the feeling that my children are in jeopardy. The next morning, I called again, hoping to catch Britt between classes.

I was relieved to hear Britt's cheerful voice on the other end of the line.

"Oh, hi Mom! Yes, I'm here for just a minute exchanging one set of books for another. My friend Louisa helped me with my French this morning at breakfast; she says I have a wonderful French accent. Anyway the paper is in and I have to go to Art 100 in about two minutes. I'll call you over the weekend. I love you." Often when I made these follow-up calls, Britt didn't seem to remember that she had been distraught the day before.

Still the distress calls continued. One evening in November, Britt called crying. This was the fifth night in a row that I had talked to her. In each conversation, she had been increasingly upset, though she was unable to say specifically what the problem was. I had become so worried I couldn't concentrate on my own work. Finally, I asked, "Britt, do you want me to fly out for the weekend? I could give you moral support."

"I don't want you to do that," Britt replied without hesitation. Evidently, the thought of her mother coming to campus was worse than having to go it alone. After I threatened to make a trip, Britt was less upset during our phone conversations, but she was still anxious. She told me that she thought she was failing, but when I asked about her grades, she reported mostly "B's" with occasional "A's" and "C's." I tried to assure her that these grades were great, but she continued to feel that everyone was smarter and more academically successful than she was.

Much later she came to realize that the East Coast culture was partly to blame. As she said over Christmas break, "My grades were just as good as most other students', but I'm just more laid back about things." Britt also discovered that two of her best friends had tied for the highest grades in the freshman class. Britt was measuring herself against a biased sample.

Because of her level of homesickness, we decided that this first year she should come home for Thanksgiving break even if it took almost an entire day of travel each way. Britt negotiated the airports including a change of planes at O'Hare without difficulty, and she was thrilled to be home for Thanksgiving. My cooking had never tasted so good, her cat was particularly cuddly, and her bed was twice as comfortable as she remembered it. When Sunday came, I knew she was sad to leave, but I

> " I awoke with the knot in my stomach that is associated with the feeling that my children are in jeopardy. "

assured her Christmas break was only three weeks away, and then she would be home for a whole month.

Christmas was close, but so were finals—Britt's first college finals. Her anxiety ebbed and flowed as she wrote her final papers and began studying for exams. On Tuesday night, Britt called to say that she had taken two of her finals, and they had gone pretty well. I breathed a sigh of relief and decided that I had the energy to bake another batch of Christmas cookies before I went to bed.

On Thursday night the phone rang again. This time a tearful Britt greeted me on the other end of the line.

"I have so much to do I don't know how I can get it all done. And I'm so tired. I've been studying all day for my Art History test, and I have a take-home final to finish for English before I go to the airport. And

The Source for Visual-Spatial Disorders

Julie: Parenting a College Student, *continued*

I have to pack and the airporter is coming at 10:30 so I barely have time to get my stuff after the test." Britt said without stopping.

"Okay," I said. "If you have been studying all day, then put it aside and finish the take-home test. Then pack. But most importantly, get some sleep. You can do it; it's just one more day. You can relax on the plane and then you'll be home. Now I'm going to hang up so you can finish. I'll see you tomorrow. I'll meet you at the gate."

"Thanks, Mom. I love you."

The next day I was at home wrapping Christmas presents and preparing a celebratory dinner when the phone rang.

"Mom," Britt sobbed. "The plane left without me. They can't get me home until after Christmas, and the dorms close tonight. What am I going to do? They say it's my fault so they won't do anything," Britt sobbed.

"Okay, let me see if we can work on this problem from here. I think John knows someone at the airline. Call me back in 10 or 15 minutes."

I called John and asked him to make some calls while I kept the line open for Britt to call me back. As I waited, I was filled with self-recrimination again—I should never have let her go so far from home; Britt has trouble with time and numbers; I should have known this would happen. Timed passed, I paced.

The phone rang. It was John. He had reached his friend at

> **"** *They can't get me home until after Christmas, and the dorms close tonight.* **"**

USAir who would try to figure out some way for us to get Britt home—before Christmas.

Time passed. The knot in my stomach tightened. Finally the phone rang again. "Mom, they're working on it. I'll call you back when I know more," Britt said in her confident, take-control voice.

Still full of anxiety, I wiped fingerprints from the refrigerator and straightened pictures. I jumped when the phone rang.

"Mom, it's okay." Britt must have heard my worried-mother's voice. "Here's what happened. A manager saw me sitting here and said 'You've been here for a long time. What's wrong?' So I told him and he went into the record and found out that they had changed the flight. It wasn't my fault at all. The stupid woman at the check-in counter didn't even try to find out what was wrong. Anyway, they're going to put me up at a hotel, and they found a first-class seat for me at 7 a.m. tomorrow morning. I'll have to change planes in Philadelphia and again in Pittsburgh, but I don't mind. I'm coming home!"

"Oh, Britt, I'm so relieved. What are you going to do now?"

"I'm going to the hotel, I'm going to order room service, and then I'm going to watch television and sleep. I'll call you when I get to the hotel," Britt replied.

I sat down at the kitchen table, hugged the cat, and cried with relief. John's friend may have helped get things moving at the airport, but Britt had managed the situation. Britt had not made a mistake and she had handled a problem that would fill any holiday traveler with anxiety. The welcome home dinner could wait until tomorrow night.

ঔ ঔ ঔ

The Source for Visual-Spatial Disorders

Britt: Finding My Place

After my initial homesickness freshman year, I began to settle into a comfortable routine. One of the things that had appealed to me about Smith was that the dorms were more like large homes than dormitories. Most of the students lived in a "house" for their entire four years, so Northrop House was a good mix of students from each class.

This mix of students set the tenor for the dining room—a bright yet cozy room with hardwood floors, long wooden tables and chairs, and floor to ceiling windows. At precisely 6 p.m. each evening, the double doors would swing open and a horde of women would thunder down the wooden staircase and into the dining room.

Once a week, we had dinners with candlelight. Platters of roast beef and steaming bowls of vegetables and mashed potatoes were passed around each table. This provided us an opportunity to invite guests—professors, deans, or visiting parents. We all looked forward to Thursday night dinners, but I especially appreciated them because they provided the opportunity for lively intellectual discussion. While we had good dinner discussions every night, we lingered longer over coffee on Thursday evenings. Christine, who did not live on our floor, often joined our table, and through these conversations, Christine became one of my best friends.

Friday afternoon tea, another aspect of "house" life, was a tradition at Smith that I enjoyed. All week long students were extremely focused on their studies. Like caterpillars coming

> **" *I had taken my place in a department at Smith; I had found an academic home.* "**

out of our cocoons, we gathered in the living room at 4 p.m. every Friday. Several steaming pots of tea, china cups, cookies, cheeses, crackers, fruit, and sweet breads were set out. Friday tea provided a chance to talk with people I might not normally talk to. I marveled at the fact that we took time from our busy study schedules to relax and socialize.

Everyone else I knew had no problem choosing a major. It seemed as if they had come to school with clear notions of what they should be doing. I, on the other hand, had trouble deciding on a major. But by the end of my sophomore year, I had to declare.

I loved art history and considered that, but I discovered that I would have to take studio art and all I could think of was the difficulty I had had making the simplest drawings in my high school art classes. I loved literature, but my freshman English class had poisoned my opinion of the English department. I also figured that an English department would be entirely too picky about grammar. I loved my history courses, but history had a stiff language requirement and there wasn't enough academic support for me to continue with French. Finally, I settled on American Studies that allowed me to take courses from history, literature, art, and government.

One Thursday night at dinner, I made the announcement of my major to my table, and it was greeted with cheers. I told them that my only fear was finding an advisor. The next day my friend Christine walked with me to the American Studies offices to encourage me and to make sure I didn't back out. I meekly knocked on the door of Professor Horowitz, the chair of the department, and asked if he would be my advisor. He agreed. I signed the papers and became an American Studies major. I had taken my place in a department at Smith; I had found an academic home.

Having grounding in a department gave me guidelines for making decisions about my academic career. It gave me the framework I had been looking for. I'd finally found professors that I enjoyed working with

Britt: Finding My Place, continued

and managed to find classes that I was comfortable with. I was enjoying learning, and I was more able to become a part of the Smith community.

At the beginning of my senior year, my sense of belonging increased when I was appointed to the Board of Trustees as one of the student representatives. I enjoyed the meetings because they gave me a broader view of the college. The President of the Board of Trustees asked us to think about the whole community—the students, faculty, staff, and administrators. At the same time, I was on the Dean's Advisory Board, so I had been privy to many issues affecting the college, especially the recent passage of the Americans with Disabilities Act. Because the college was ready to embark on major renovation projects that would meet ADA standards, I was excited to be playing a role in the future of Smith.

After three years in Northrop House, I was ready for a change. Choni, Louisa, and I decided to try for one of the senior apartments on campus. There were only about 15 units available. They were decided through a lottery system. We were lucky and managed to draw the very last winning number. Louisa came back from her junior year in Florence and we moved in. We threw together some odds and ends to furnish the kitchen and bought a TV that worked well enough for our purposes. Mom came to Northampton to help us move and see the apartment. She got us a small grill for the patio. Yes, this would be a good year, and we all had high hopes.

The time had come to choose classes for the last semester of my senior year. I had not ventured out into any other subjects aside from History, Art History, English and Government. Smith did not have any math requirements, and I knew that I'd have trouble with science or math. But this was my senior year, and I wanted to try something new. I chose an Introduction to Psychology class. I thought that this would be a non-stressful way to branch out into the sciences. The class was a non-lecture class and relied solely on tests from the textbook. Plus I had two senior seminars and a large research project that were going to take up a lot of time and energy. I headed into the last semester with optimism.

The senior American Studies seminar I had signed up for was entitled "Writing Biography." We studied different methods of research and different forms of portraying points-of-view and how the author influenced the reader's perception of the person they were writing about. For the final project, we had to pick a person and write a biography about him or her.

I took a trip to the college archives to check out the Sophia Smith collection. I chose Sara Payson Willis, pen name Fanny Fern. She was the first female newspaper journalist in the United States. Her life and writing seemed like an interesting study in the mid to late 19th century culture, and there was plenty of information about her in the archives. I spent hours poring over letters and journals, my hands covered with felt gloves to protect the aging materials.

> *"The one class I expected to whiz through turned out to be a nightmare."*

Psychology was a different story; as much as I enjoyed my biography class, I dreaded psychology. The one class I expected to whiz through turned out to be a nightmare. We met twice a week to take chapter exams. These tests were divided into two parts. The first part was multiple choice consisting of 30 to 40 questions. The second half was oral. A proctor asked each student a series of questions regarding the chapter.

I studied for the chapter tests, reviewed the material in the book, and even went over the chapter quiz provided in the book. I was ready, or so I thought, until I arrived and took the first multiple-choice quiz. I looked at the questions and chose what I thought to be the right answers. I handed the test sheet to the professor who, with an unwieldy red pen, began marking all of the wrong answers. I was shocked to see how many I'd missed.

When I got back to my room, I went though the quiz and compared my wrong answers to the correct ones. I immediately noticed that the answers I'd marked as correct differed only slightly, usually by only a single word, to that of the right one. We had to keep retaking the test until we got enough right answers to pass before moving on to the oral section. I took the test six times before I moved on to the oral section. Once I got to the oral part, I passed it in one session—not missing a single question. This same pattern repeated itself through the entire semester.

After having trouble with the quizzes in Psychology, I went to the disabilities coordinator to see what could be done. I requested that I receive the exams aloud. I thought that hearing the questions read would help me make the distinctions necessary to answer the multiple-choice questions. The disabilities coordinator told me to ask the professor. When I approached the professor, she told me that the tests were standard, and they wouldn't change the format for any reason. I kept taking the tests and retaking the tests with the same results. With the pressures of midterms and research due for the biography project, I felt that I was beginning to sink and fast. I went back to the disabilities coordinator and asked again if there was anything she could do. The answer again was no. I was in danger of failing my Psychology class.

The senior class dean was notified, and I was called into her office. When I went in to talk with her, I expected her to tell me that she would try to resolve the situation. Instead, the first words out of her mouth were, "Why did you choose a class you knew you couldn't handle? You have to choose your battles." I was shocked and replied that I expected to handle a freshman-level psychology class. This was a battle that should be fought. I told her that I chose this class because I thought it matched my strengths as well as allowed me to branch out academically.

I called Mom from the basement phone in Seelye Hall and could not hold back the tears.

Mom reassured me that she would try to do something. She called the next day to say that she had called the Office of Civil Rights in Boston and was waiting for a reply.

The civil rights officer at the college was notified of the complaint, and the Psychology Department acquiesced just a week before final exams. The psychology professor called me into her office and told me that the department had a meeting and would reluctantly allow the change in testing. But unfortunately, time ran out in the semester before I had a chance to take all of the chapter tests. I eked out a "C-," enough to pass and graduate.

The Psychology exam was the last one I ever took at Smith, and when I walked out of the Science building, I almost expected music to trumpet from the skies. Instead, I rushed back to the apartment where Choni and Louisa were waiting to start the celebration.

ಙ ಙ ಙ

> *" I should be expected to handle a freshman-level psychology class. This was a battle that should be fought. "*

Julie: Parent Advocacy

As Britt moved into her sophomore and junior years, she became increasingly independent. She arranged her own transportation to and from the Hartford airport, chose American Studies as her major, and joined various campus organizations. During a parents' weekend visit, I met her friends—Louisa, Choni, Bronnie, Christine, and many others. She had taken responsibility for her disability, arranging accommodations and sought help as she needed it. All was well as her graduation approached.

One night during the spring of her senior year, Britt called to say that she had not been able to pass the tests in her Psychology course. The tests were impossible for her. She knew the material, but the format called for fine distinctions in wording. I suggested she go to the school's disabilities coordinator to seek accommodations, probably a different test format.

Britt started to cry. "Mother, I've tried. The professor says that I have to do it this way. She doesn't understand that I know it; she thinks I'm trying to find an easy way out."

"Britt, we know you are hard working and smart. How about trying the disabilities person one more time?"

The next morning my phone rang at work. It was Britt sobbing again. The disabilities specialist had sent Britt back to the professor. "Okay," I said, "Let me fax you a copy of the law and a couple of court cases on this topic. You are entitled to accommodation."

The professor refused to discuss the matter or look at the cases Britt was presenting to her. Britt was so upset that her other courses, which she loved and had been successful in, were suffering. If Britt did not pass this course, she would not graduate with her class in May.

Through college, I had let Britt be her own advocate, but now it was time to get involved. I was convinced that her rights were being violated, and I could not let her fail through no fault of her own.

I called the Office of Civil Rights in Boston and told the person I spoke with that I wanted to file a grievance against the college and against the professor. I explained that we did not want to do this, but we felt we had no other choice. The officer, both professional and kind, said that it certainly sounded as if we had a grievance. I explained that time was an issue and that all of Britt's documentation was on file at the college. She suggested working directly with the civil rights officer at the college to help reach a speedier resolution. She said that if we were not entirely satisfied with the outcome, I should call her back, and she would help us file the official grievance.

Within the hour, the college civil rights officer called to say that she would be in touch with Britt immediately to reassure her. Then she would call the professor to explain why she must provide an alternative test format for Britt. The officer said she agreed with us completely; Britt's rights had been violated and she intended to take care of it—now.

My part was done, and Britt had been vindicated. I had worried that I had been an overly protective, meddling mother. Clearly, I had not. Britt was an adult, yes, but sometimes she still needed an advocate. I was glad I had stood up for what was right.

༄ ༄ ༄

> **"** *Through college, I had let Britt be her own advocate, but now it was time to get involved.* **"**

Graduation

Carol

Britt called in early May. Smiling at the contagious enthusiasm in her voice, I listened to her describe exciting upcoming events.

Britt was graduating. Of course she was! Everyone who really knew her expected that she would. But still, this moment deserved a bit of reflection. She had, after all, traveled a long way since fourth grade.

She taught me so much—that intelligence, even brilliance, can dwell side-by-side with invisible obstacles and that these obstacles can be overcome. She had not needed help from me for quite some time. But how difficult it was for some to see intelligence because they focused on the obstacles—on what Britt and other students could not do rather than on what they could do. Britt had already helped some of her teachers understand, but there was still a great deal of work to be done.

There was much more to celebrate than the diploma, although that was reason enough. Britt had become an independent adult who was ready for life's challenges. Britt, Julie, and I joyfully concluded our unusual journey as three adult friends and traveling partners, reflecting back to the time when only faith gave us the courage to begin. The time had come to celebrate.

Julie

We arrived in Northampton for graduation on a warm and humid May evening. Britt was jubilant—she had turned in her last papers, she had taken her last tests, and she had passed the psychology course. She could look forward to a weekend of celebrating. For me the weekend went by in a blur of images—the songs and lanterns that covered the campus on Friday night; the hugs and congratulations Britt received from the Trustees at the reception in the Art Gallery; the quiet of Britt's favorite place in the now deserted library; the sea of young women in white on Ivy Day; and finally Britt's joy, all of our joy, as Mary Maples Dunn handed Britt the Smith College diploma. There was much to celebrate.

Britt

The week after finals and before graduation passed in a flurry of activity. There were tons of things to pack—books, clothes, and my computer.

Choni's mom arrived from Sri Lanka, her first time in the United States, and Thursday night she whipped up a Sri Lankan feast filling the entire apartment with the tantalizing aromas of exotic curries and spices. The heaping bowls of saffron rice, coconut curried chicken, deviled beef, and rice custard tasted wonderful. As we finished our dinner, my family arrived—Mom and John as well as my brother John and stepsister Sarah.

Friday was filled with awards ceremonies, receptions, and speakers. I gave everyone a tour of the campus and introduced my family to my advisor and other professors. After dinner on Friday night, the campus was lit up with hundreds of shimmering Japanese lanterns along the winding walkways. Musical groups entertained graduates, parents, and alumnae who had gathered for the weekend's festivities.

Early Saturday morning, Choni, Louisa, and I readied ourselves for the Ivy Day procession. I put on my white silk blouse and skirt, joining Choni and Louisa for a quick set of pictures. As we stood smiling for the camera, a wave of sadness swept over me as I realized how kind and supportive they had been. I also knew that I was unlikely to ever have such good friends again. We headed out to join the other seniors—all in white—lining up in the quad. We were amazed as we watched the alumnae march through the quad. Classes from as far back as 1918 were present,

Graduation, *continued*

and we all felt the powerful sense of tradition that these intelligent, determined women represented.

On Graduation Day, we wore our black caps and gowns and took our places among the other graduating seniors. I caught a glimpse of Christine and waved as I found my place. The line of seniors stretched across the entire quad, and we had a difficult time containing our excitement as we waited for the ceremony to start. A loud blast from bagpipes and tapping of drums marked the beginning of the procession, and by house, we all marched in and took our seats. Ruth Bader Ginsberg received an honorary degree, and our commencement speaker, congresswoman Jane Harmon, a Smith alumna, delivered her speech and wished us all well. Then it was time to receive our diplomas. I watched happily and cheered as Choni and Christine went up to receive their diplomas. I heard my name called and headed up to accept a diploma and gleefully shake hands with Mary Maples Dunn. I returned to my seat and looked down at the diploma I'd been handed. "Rachael Brown, summa cum laude," I read in amazement, and thought maybe I should keep this one. Then I remembered that in Smith tradition, each woman receives someone else's diploma during the ceremony to remind the seniors that they are all connected to the Smith community and bound together as Smith College graduates.

After the actual ceremony, 600 seniors rushed to the large grassy quad where we raucously passed diplomas until all seniors had received their own. One by one seniors left the circle. I waited, checking each diploma. Just when I thought that I'd never find mine, I saw Christine across the quad, waving my diploma. She walked quickly to me and handed me the diploma. We gave each other a triumphant hug, and I said, "Christine, we made it!"

ಙ ಙ ಙ

Part II: Helping Students with Visual-Spatial Problems

Students with visual-spatial problems may have difficulties in a variety of academic related areas, including:

- Language of Space
- Grammar and Language Mechanics
- Time
- Mapping
- Handwriting and Drawing
- Distance, Weight, and Size
- Mathematics
    - Number and Place Value
    - Operations
    - Story Problems
- Writing

Background

Frame of Reference

Imagery

Language

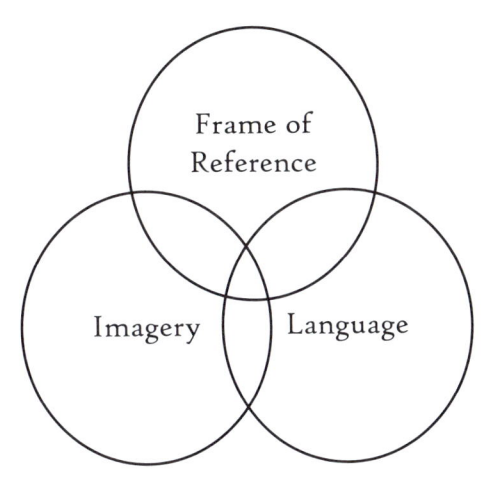

Spatial Tasks Require Three Vital Learning Processes

While Britt's strong language was a key, she needed all three basic learning processes (frame of reference, imagery, and language). This often meant developing images or frames of reference before she could use her language to describe or discuss the information. All these steps require the frame of reference first.

Frame of Reference. Britt cannot understand "Go north for three miles" unless she has a frame of reference that includes south, east, and west. Moreover, she must be able to look at the street and know or be able to figure out north. All reference systems are built upon comparison and contrast. Students with spatial disorders frequently have flawed or missing frames of reference.

Imagery. Britt's images are her retrieval system of what she has seen, heard, smelled, tasted, and felt. They may even be for movements such as handwriting. Images are the "substance" of memory, but they are also constructions. All of us create images; we imagine objects, people, or events that never existed or even those that never could. Imagery helps Britt problem solve. She images previous experiences and uses them to solve a present challenge by comparing, measuring, and weighing information. All these steps use the frame of reference. She must have that first.

Language. Britt uses self-talk as a tool to keep herself on target as she applies the frame of reference to the images of information to guide herself toward goals. While the goal is often imaged, self-talk is what keeps that goal in front of her as she figures out what to do. For example: *"Let's see, I need to be at the medical building by 11:30. I have to stop and get gas. That will take at least 15 minutes. Driving time is 30 minutes. If I am picking up Helen, I will have to leave by 10:30."*

The Source for Visual-Spatial Disorders   Copyright © 2002 LinguiSystems, Inc.

Getting Started

Begin at the beginning. Commit yourself to looking at each student with spatial problems as a unique individual with specific needs. Formal assessment may be advisable. Your student may or may not be like Britt. The absolutely critical first step is recognizing and understanding your student's processing strengths and weaknesses. How can you do this?

◆ **Take a careful history of physical and social development, and academic progress.**

Did the student . . . ?

☐ walk early
☐ climb
☐ easily tie her shoes
☐ ride a bike

Does the student . . . ?

☐ play on a soccer team
☐ find her way to new locations
☐ keep track of her possessions
☐ like to draw
☐ enjoy jigsaw puzzles

Is the student . . . ?

☐ fearful

◆ **Observe the student.**

Does the student . . . ?

☐ follow instructions
☐ interact with his teacher and peers
☐ become lost in his own neighborhood or school
☐ work independently

◆ **Identify academic skills.**

Did the student . . . ?

☐ learn to read at the usual age

Does the student . . . ?

☐ read for pleasure
☐ spell appropriate to grade level
☐ write efficiently at a grade-appropriate level

Can the student . . . ?

☐ add, subtract, multiply, and divide
☐ borrow, carry, do fractions and story problems
☐ tell time
☐ understand money

The Source for Visual-Spatial Disorders

Getting Started, *continued*

◆ **Look for patterns. Know the nature and extent of each skill.**

Let the learning pattern emerge from the evaluative process. One informal but effective way to sort out where to begin work is to make "Can Do" and "Cannot Do" lists. Be thorough.

Can Do	Can't Do
Read textbooks	Add 2-digit numbers
Count to a thousand	Find specific pages in books
Ride a bicycle	Draw

◆ **Ask questions and make observations to define needs.**

- ☐ Can the student identify specific problems?
- ☐ Are the problems long-standing or have they recently appeared?
- ☐ Are the student's problems limited to specific skills such as reading?
- ☐ Track the instances when problems occur and the frequency of such instances.

◆ **Develop a mindset for instruction that works.**

After identifying your student's patterns of strengths and problems, you are ready to begin instruction.

Instructing a student with a spatial disorder can be frustrating to the point of bewilderment or it can be joyful and fulfilling. Britt could not "undo" her disability so her teachers and mother had to understand the processes she needed to use to succeed.

Your student may have other stumbling blocks that Britt did not have. Unlike Britt, your student might need to develop adequate language before he can use it to guide or transfer the learning. Or he may have motor-planning problems that interfere with writing. Inattention, limited experiences, or various medical conditions can result in meager or generalized memory traces. Know your student and provide for his particular set of stumbling blocks.

The Language of Space

> *For students who do not use comparative or relational words when talking or writing*

Students with spatial disorders often have difficulty with the spatial features of language. Britt did not because of some very early language games she played. Students need to learn words that show relationships before they can use self-talk in math or other tasks. Some examples are listed below.

- **Place**: there, here, where, wherever, above, below, beside, into, over, under
- **Direction**: right, left, north, south, east, west, perpendicular
- **Contrast**: although, while, even though
- **Cause**: so, so that, as, since, because
- **Condition**: if, when, provided that, in case, assuming that
- **Time**: when, whenever, while, once, before, after, since, until, as long as, as soon as

Students have usually heard these common words over and over without gaining control over them and the relationships they represent. Therefore, building a frame-of-reference and developing firm images for the words need to be linked. The student who acts out the relationship while hearing and using the words is honing an image for the words. Students must have feedback—immediate feedback—so the words and images are linked.

One way to build a vocabulary of spatially descriptive words is to play the "Say Where" game. This game is described on pages 167-169.

The Language of Space, continued

Frame of Reference:
The visual comparison of the block arrangements provides a reference.

Imagery:
Imaging the meaning of the directions rehearses and reinforces imaging.

Language:
The need for words that describe placement builds spatial vocabulary.

"Say Where" Game*

Purpose: to build the vocabulary for spatial relationships, link that vocabulary to images, and rehearse

Each player needs:
- a folder
- different shaped blocks
- a rectangle of paper that is used as a base

Use the folder as a dividing screen between players. Each player has the same number and shapes of blocks and a paper rectangle. The person who is "It" arranges his blocks on the rectangle and then describes the placement of each block. Other players follow these directions. Then they take down the divider and compare the final results.

A description of three people playing this game is found on the following pages.

* Stockdale, C. and C. Possin. *The Source for Solving Reading Problems*, East Moline, IL: LinguiSystems, Inc., 2001, pg. 147.

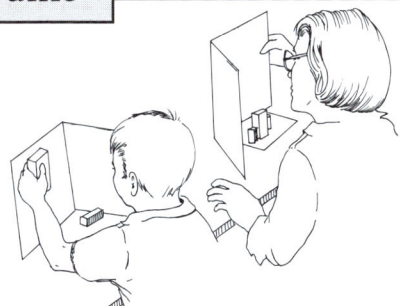

The Source for Visual-Spatial Disorders

The Language of Space, continued

1

"Please put the square in the center of the paper."

Eric is "It" so he gives directions.

"Center? That's middle. I hope this is the middle."

Fred has a visual-spatial disorder.

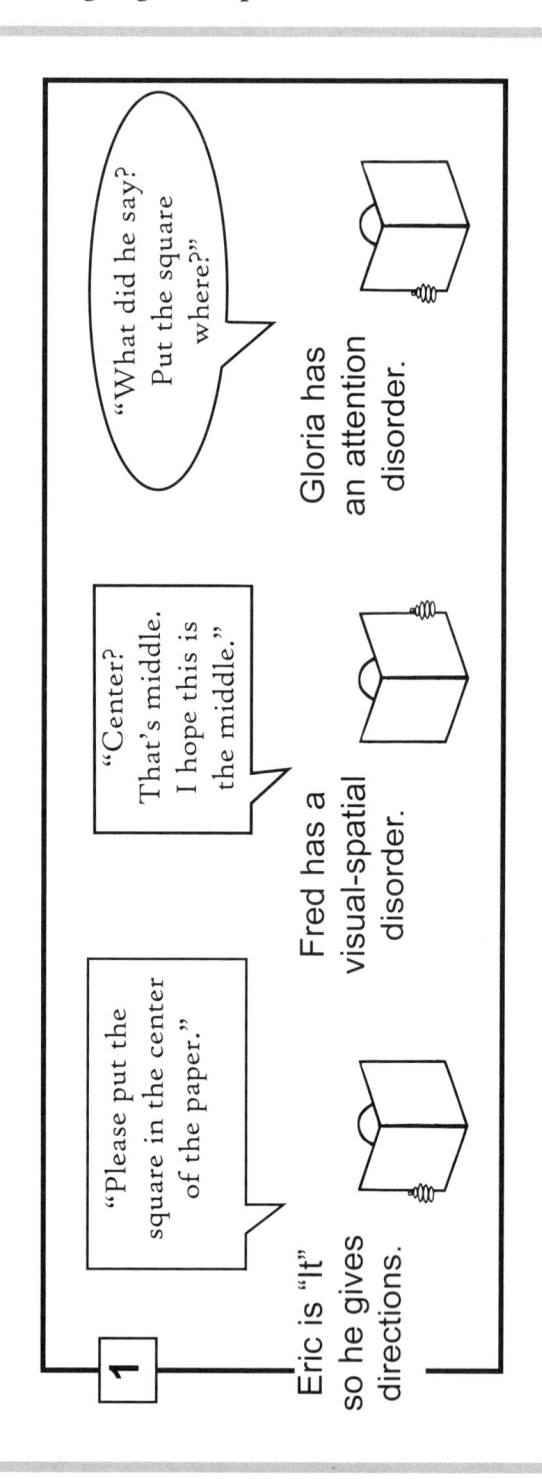

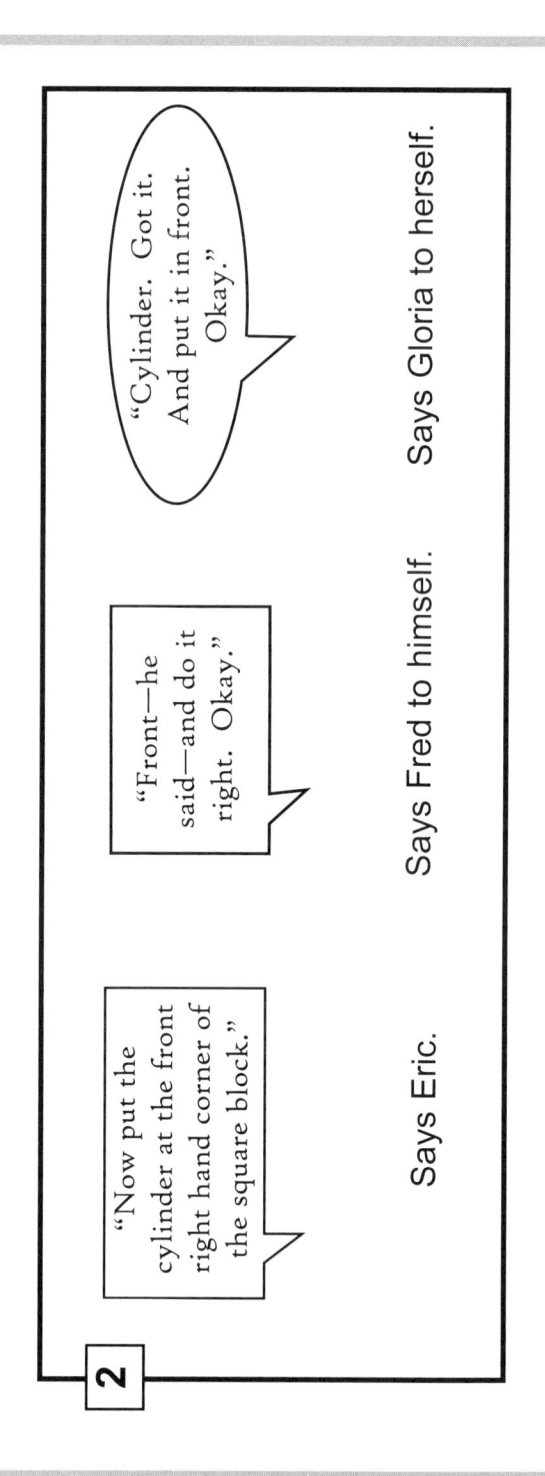

"What did he say? Put the square where?"

Gloria has an attention disorder.

2

"Now put the cylinder at the front right hand corner of the square block."

Says Eric.

"Front—he said—and do it right. Okay."

Says Fred to himself.

"Cylinder. Got it. And put it in front. Okay."

Says Gloria to herself.

Frame of Reference: Live action with real blocks provides a tangible, visible reference for words that describe shapes and placement.

Imagery: The players must transfer the oral language into images in order to select and position blocks to represent the oral instruction.

The Source for Visual-Spatial Disorders

The Language of Space, *continued*

3

"Lay the rectangular block flat on top of the cylinder."

Directs Eric.

"Rectangle flat on cylinder—which way?"

Says Fred—puzzled.

"Rectangle on cylinder. Got it. There."

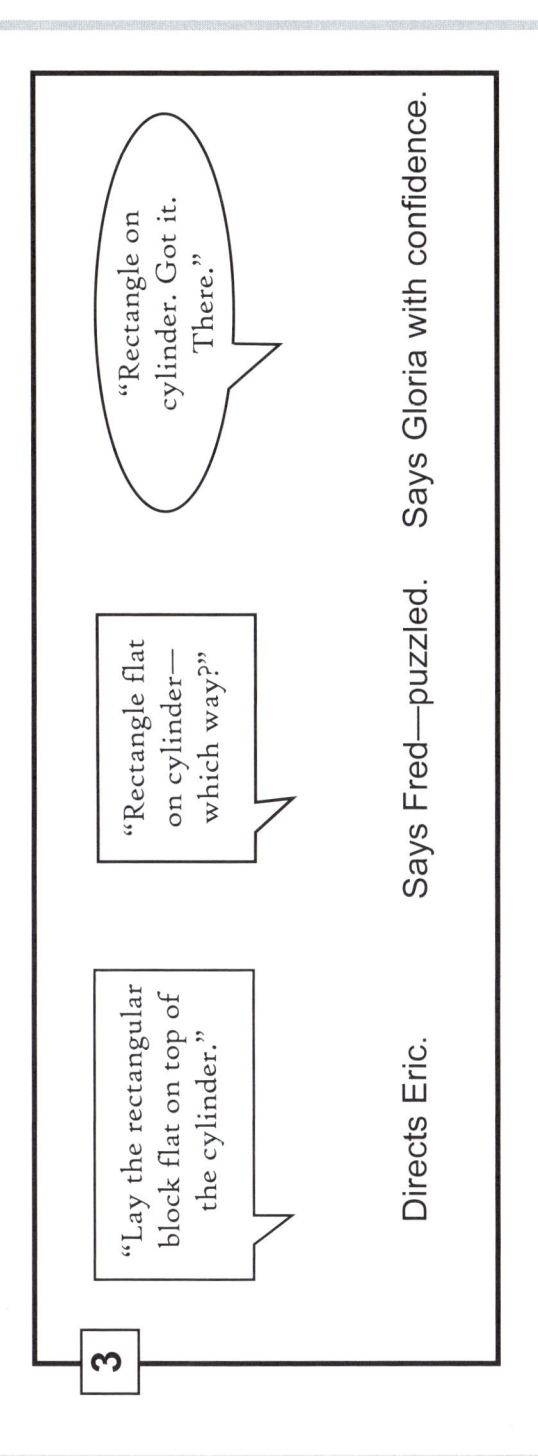

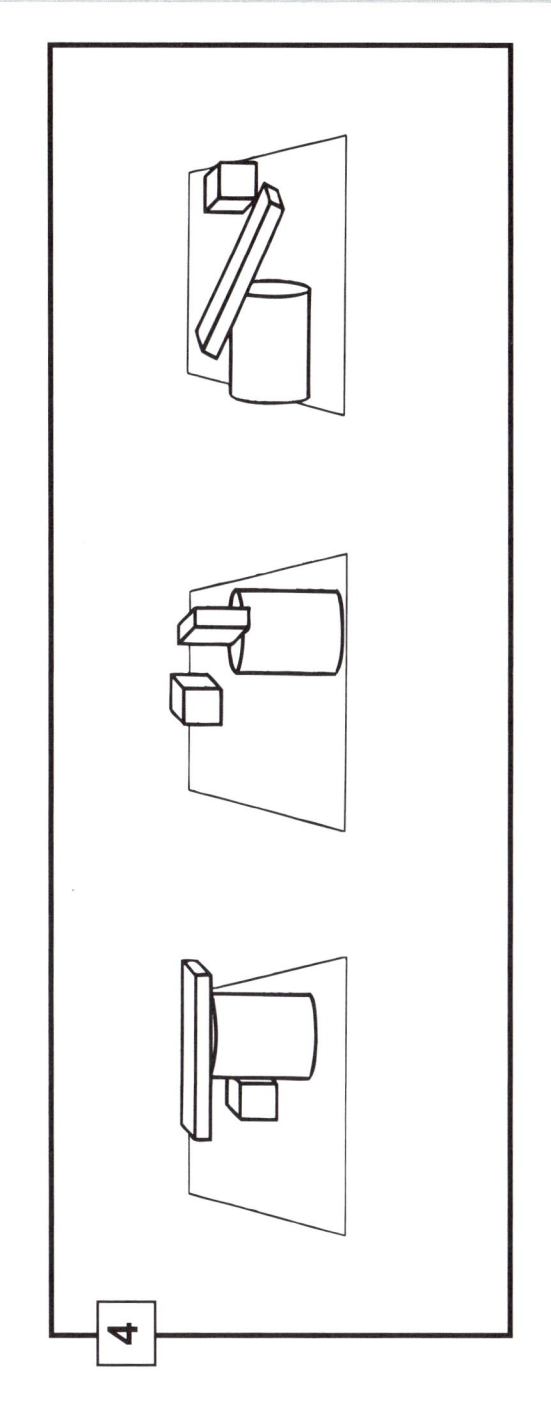

Says Gloria with confidence.

4

By the third round, both Fred and Gloria produced matches to Eric's directions.

Language:
Whoever is "It" must carefully shape specific commands.
The players must recall and retain the command while they carry out the designated action. They get feedback that tells them what was correct and what they need to change.

The Source for Visual-Spatial Disorders 169 Copyright © 2002 LinguiSystems, Inc.

Grammar and Language Mechanics

> *For students who write incomplete sentences and who cannot edit*

Frame of Reference:
Arranging the Word Shapes creates an awareness of word function in sentences.

Imagery:
The colors, shapes, and sentence patterns help students internalize language configuration.

Language:
Students first must be able to control the form of language to use it as a tool for transferring meaning.

Students may speak correctly because they imitate the patterns of their families but still not understand standard language patterns to guide them during reading and written composition. This is frequently an area that is very difficult for students with spatial disorders. Language patterns are spatial arrangements. Learning labels such as *verb*, *adjective*, or *preposition* may or may not be possible. Either way, this labeling seldom leads to accurate composition or editing skills for these students. However, most students benefit from working with a visual representation of language patterns. Word Shapes were developed to serve this function. They are easy and fun. Most students quickly "feel" common language patterns after constructing them with the plastic shapes. Examples of how to represent several language patterns with Word Shapes are illustrated below.

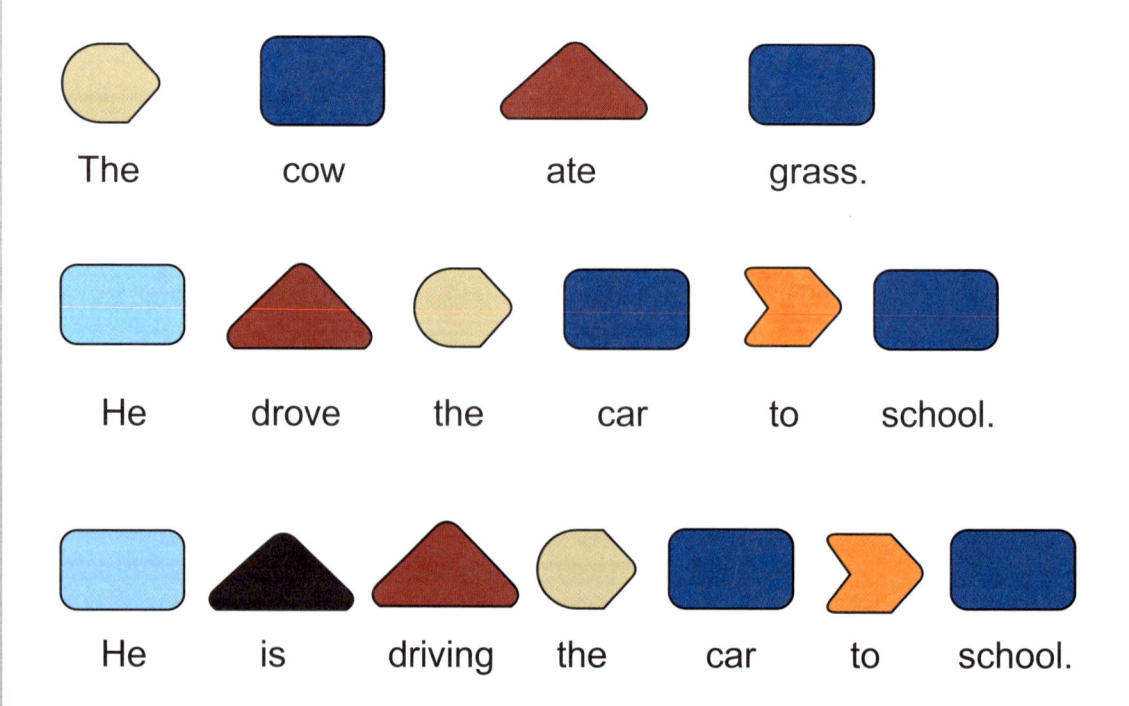

The   cow   ate   grass.

He   drove   the   car   to   school.

He   is   driving   the   car   to   school.

See page 171 for a key listing each colored shape and the part of speech it represents.

Word Shapes

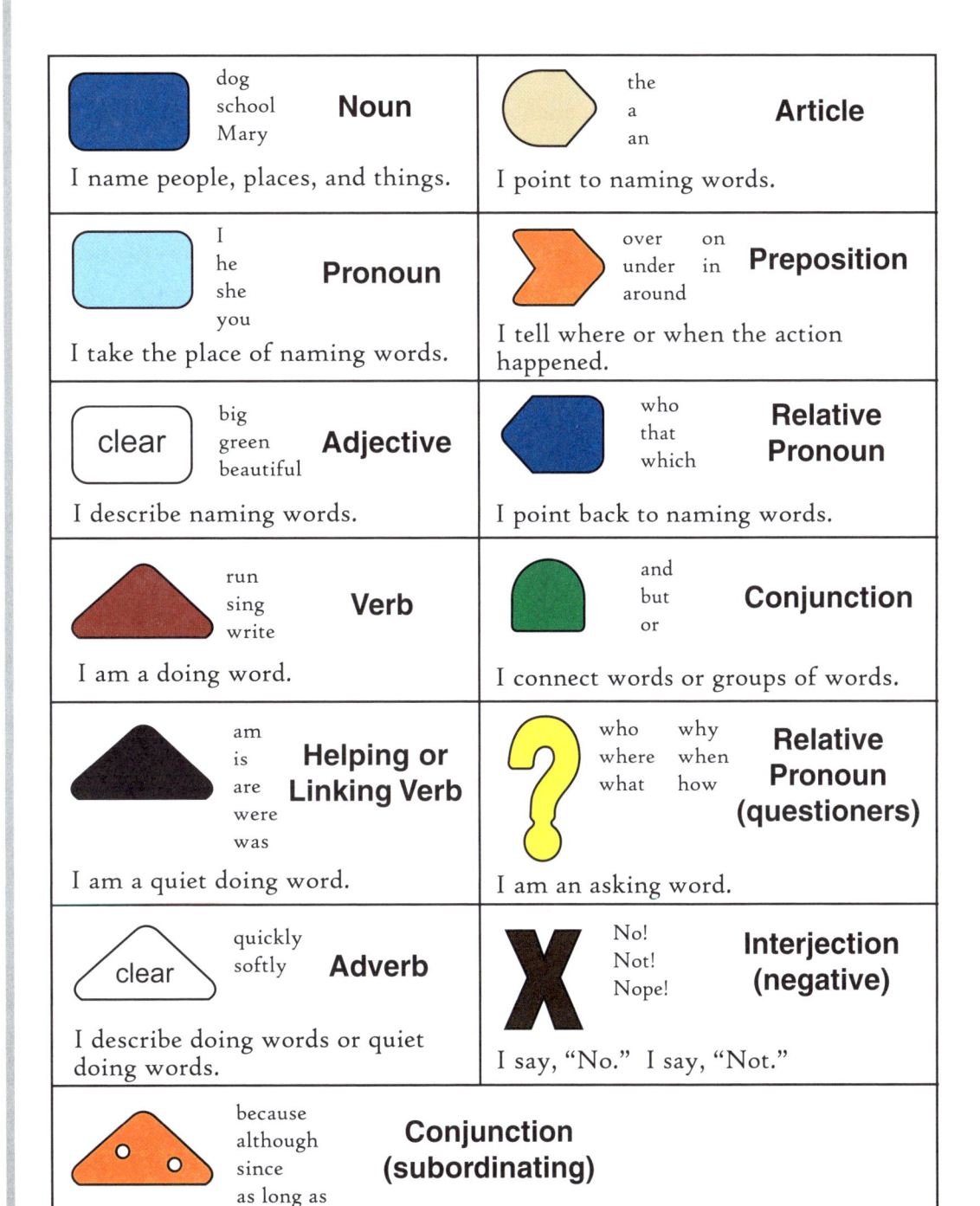

▇	dog school Mary	**Noun**	the a an	**Article**
	I name people, places, and things.		I point to naming words.	
▇	I he she you	**Pronoun**	over on under in around	**Preposition**
	I take the place of naming words.		I tell where or when the action happened.	
clear	big green beautiful	**Adjective**	who that which	**Relative Pronoun**
	I describe naming words.		I point back to naming words.	
▲	run sing write	**Verb**	and but or	**Conjunction**
	I am a doing word.		I connect words or groups of words.	
▲	am is are were was	**Helping or Linking Verb**	who why where when what how	**Relative Pronoun (questioners)**
	I am a quiet doing word.		I am an asking word.	
clear	quickly softly	**Adverb**	No! Not! Nope!	**Interjection (negative)**
	I describe doing words or quiet doing words.		I say, "No." I say, "Not."	
▲	because although since as long as	**Conjunction (subordinating)**		
	I hook together groups of words and show how they are related.			

Word Shapes are available through the ARK Foundation:
ARKfdn@aol.com
(253) 573-0311

* Stockdale, C. and C. Possin. *The Source for Solving Reading Problems*, East Moline, IL: LinguiSystems, Inc., 2001, pg. 137.

Learning to Tell Time

> *For students who struggle with interval, number, and systems*

Frame of Reference:
Pennies and nickels are tangible substitutes for minutes.

Imagery:
Builds on the images the students have for penny and nickel.

Language:
Self-talk keeps the frame of reference linked to images as the clock hands move.

Qualities such as abstraction, movement, relative size, numerical relationship, and interval are especially challenging for the student with visual-spatial disorder to understand and image. Yet each of these qualities is involved with reading a clock. Time is abstract. Moreover, the hands of a clock are always moving and time is always changing. Finally, the task involves the relationship of large and small hands, numerals, and the intervals between the numerals. By creating a penny clock, the student has a personal tool for translating the intangibles of time into a concrete format that she can control.

- Make a paper clock 14 inches in diameter. This is large enough to fit 60 pennies around the circumference. Add hands that move.

- Have the student count out 60 pennies, stack them in 12 piles, and then count the pennies in each pile.

- Have the student place one stack on each number on the clock, and then distribute them around the circle, filling in the spaces between numbers.

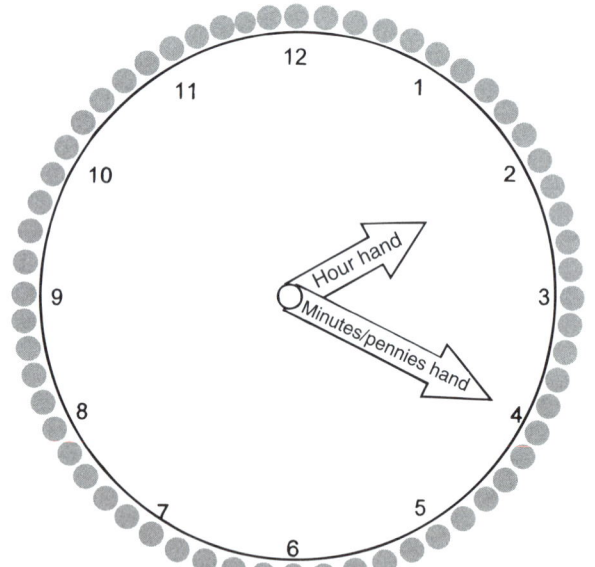

The Source for Visual-Spatial Disorders

Learning to Tell Time, continued

- Name the short hand the "hour hand" and the long hand the "minutes/pennies hand."

- Practice by saying the time in terms of pennies (e.g., 20 pennies after 2; 4 pennies before 12).

- Create various times on the penny clock from a digital watch face. Go back and forth between the two views of time.

Option:
When the student is comfortable with "penny time," stack the pennies by each number. Count them again. Substitute a nickel for each stack of five pennies. Do the time exercises with the nickels in place of the pennies.

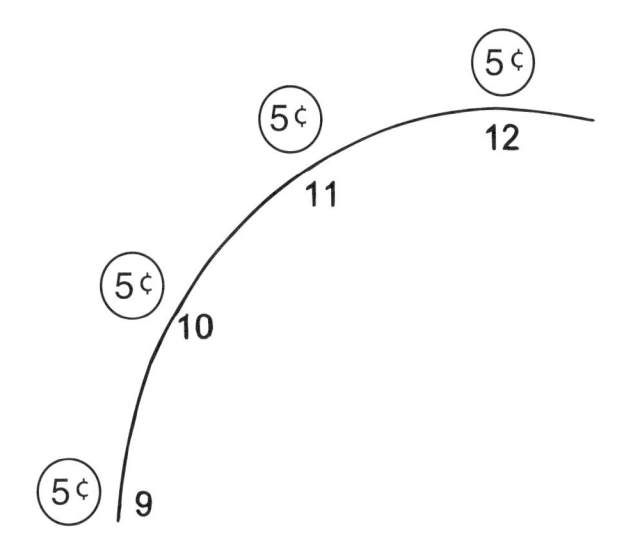

Note: Be certain to have the student describe his method for reading the clock aloud. For example, "11:15 means 15 pennies after the hour of 11. We can also say 15 minutes after 11." or "11:15 means 3 nickels (5-10-15) after the hour of 11 or 15 minutes after 11."

Mapping

> *For students who frequently get lost or cannot give directions*

Frame of Reference:
Create an internal map that links specific visual sights with direction.

Imagery:
Use images from the internal map to create routes and locate places.

Language:
Use self-talk to check the image against the frame of reference as the student guides himself with his internal map.

Finding a new destination requires both orientation to your starting point and reference to an internal map of your surroundings to plan a route. If you do not have an internal map, you need to construct one. Without a map, you are dependent upon following instructions by rote (e.g., Turn left, go past three buildings, turn right, etc.). Rote instructions are troublesome because the slightest slip-up means you are lost. Besides, they do not give you control. An internal map does give control. Students with spatial disorders frequently get lost and/or do not venture out alone. Freedom comes with learning how to construct an internal map.

Creating an Internal Map

The first step in creating an internal map is to help the student choose a location to establish his base for direction. This is the place he refers to when finding other locations. At this base, the student chooses (or makes) visible landmarks for each direction—north, south, east, and west. These landmarks need to be real three-dimensional sites that he can easily recall and describe from memory.

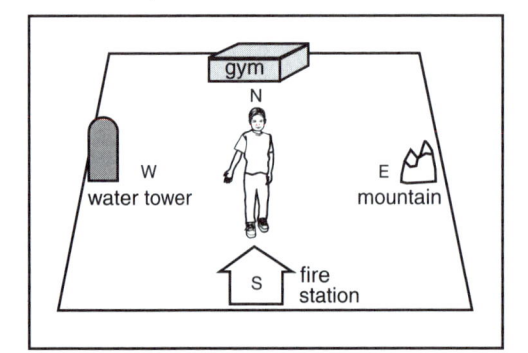

Step 1: Represent directions with 3-D objects.

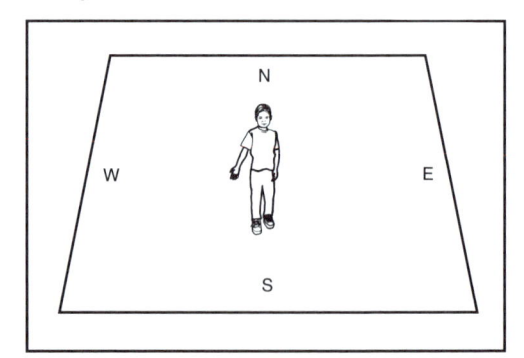

Step 2: Represent directions with letters and drawings.

The next step is to stand with the student and help him identify landmarks in visual terms such as "I am standing by the flagpole at my school facing the fire station. That fire station is south (teacher cue), so I know I am facing south." Next guide the student in looking left, assure him that he is looking east, and ask him to choose a marker. The student might say, "I am looking east and I see Mount Rainier.

Mount Rainier is east." Repeat this process for west (e.g., water tower) and north (e.g., gym.) Then practice giving directions using these sites for guides. For example, "I go west two blocks—west is toward the water tower—to get to McDonald's."

Developing the Internal Map

Students often need to extend their internal maps by anchoring these images into a larger context.

Teacher to student (with map in front of them): "If you went farther west, you would get to Puget Sound and still farther, the Pacific Ocean. Farther north is Seattle. And still farther?"
Student: "Canada!"

Cue and prompt until several markers for each direction are recorded.

Rehearsing Guidance with the Internal Map

Students need to talk their way to various sites starting with places nearby. Their talk needs to identify where they are starting from and how they know where to go.

Student: "I am in front of the flagpole and I'm going to the donut shop that is two blocks north and 1/2 block west. The gym is north so I will aim that way once I get out to the street. After two blocks, I turn west—like the water tower is west. If I kept going I would end up in the ocean!"

In this way, the student gradually builds a reliable internal map.

Handwriting and Drawing

For students who dislike writing and avoid drawing

Frame of Reference and Imagery: Through tracing activities, the individual builds a basis for understanding spatial relations as the foundation for becoming a better writer and drawer. By forming the letters and tracing basic shapes, students develop accurate images to use in a variety of learning activities.

Students with motor problems may know how they want their letters and drawings to look but still be unable to control their pencils well enough to produce that result. The Learning Window provides them with accurate motor memories from tracing the teacher's pen.

Students with visual-spatial problems can often control their pencils, but need the Learning Window to build accurate images.

Students with visual-spatial problems usually avoid drawing because they typically do not identify the distinctive features of an object, which means their drawings are often formless. They also cannot make their drawn lines follow observed shapes. Furthermore, even drawing from memory is hard because their visual memories are vague. Some, but not all, individuals with visual-spatial problems have difficulty recalling the appearance of letters and the stroke for making them.

The Learning Window is a device for teaching persons who have problems with handwriting, drawing, and working with pencil and paper. It is a transparent surface on which the learner traces and copies to form letters or pictures. A teacher guides the learner through structured exercises, giving instruction from either side of the window.

The Source for Visual-Spatial Disorders

Handwriting and Drawing, continued

Language:
Self-talk helps students write and draw as they describe how to form letters and shapes (e.g., saying "a square has four corners" helps sort shapes).

Many students need to describe the writing strokes to tell their muscles how to make them. One student described a cursive *l* as "swoop up, loop down to a valley"; *t* was "swoop up, swoop down, cross"; while *a* was "little swoop, fat tummy down."

Who needs the Learning Window?

1. Children and adults who have motor problems
   These persons lack the muscle coordination to write smoothly and accurately. They are clumsy and awkward; their writing is sloppy and meager; they tire easily with pencil and paper tasks; and they avoid writing. Their letter formation may be inconsistent (i.e., they may form a given letter in a different way each time so that they do not become efficient writers).

2. Children and adults who have visual-spatial problems
   These persons typically have trouble copying from the board or book, organizing sequences, drawing pictures, mapping, understanding quantity, applying punctuation and capitalization, understanding dimension, changing point of view, using directional language (e.g., *beside*, *beneath*, *greater than*), spacing letters and words, paragraphing, subtracting and doing other math operations, adapting to schedule changes, and telling time with understanding.

3. Children and adults with both motor problems and visual-spatial problems

Distance, Weight, and Size

For students who struggle with predicting how far, how much, or how big

Frame of Reference:
The body is the first and most basic size reference.

Imagery:
The frame of reference is very important for building useful images. Beyond creating the base, students need to rehearse using their images and checking up on the results. Their images need to be developed and confirmed in many ways.

Language:
Self-talk helps students sort and compare. It also recalls previous experiences and applies them in new situations.

Students with spatial disorders learn to avoid estimates that involve distance, weight, or size because their projections are often highly inaccurate. How far, how heavy, and how big are expressed in numbers perceived by comparison and contrast, the very skills so difficult to build for students with spatial disorders. Given appropriate experiences coupled with descriptive language, students can build a base and translate that base into images. This base and the images for that base can then be applied to new situations.

The human body is relevant for measurements. This best involves simple and direct personal physical experience. Remember, Britt needed to dip her finger an inch into the ink. Using her stained finger to measure many objects helped her "know" an inch. In the same way, a student can build a reference for 50 feet. Starting from a familiar site such as her front porch or the corner of the playground, she walks in a straight line the specified distance, stops, and marks that distance. She needs to see the entire span and walk this distance several times. This becomes her reference for 50 feet. Direct physical experience is a necessary base.

The Source for Visual-Spatial Disorders — Copyright © 2002 LinguiSystems, Inc.

Mathematics: Number and Place Value

Frame of Reference:
The student needs to link quantity to numerals and understand the number system. Overviews are very important.

Imagery:
Numerals must trigger images of quantity and quantity must trigger images of numerals.

Language:
Students must use their own self-talk to describe numeral meaning, place value, and the number system.

Self-talk is the vehicle for transfer.

Students with visual-spatial problems frequently have problems with mathematics although those problems differ widely. Number, as opposed to counting, is challenging because quantity is built on comparison. How much is 23 and how does that differ from 32? Our entire place value system derives meaning from location which means place value has inherent challenges centered in spatial relationships.

Manipulatives for creating various number combinations can be helpful if the student makes the number in beads, explains the quantity in words, and writes the numerals. He then needs to do the opposite—write numerals, read them, and build the quantity using beads.

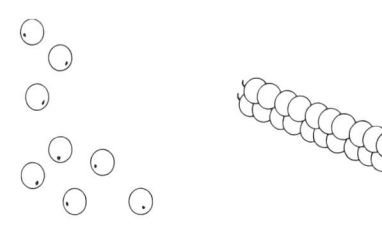

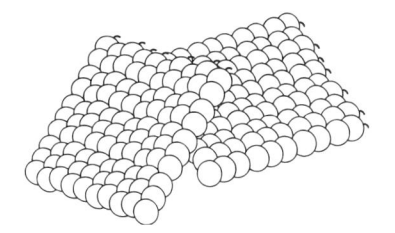

Constructing and dismantling numbers is very helpful. Show that 428 is actually 8 + 20 + 400.

		8
	2	0
4	0	0
4	2	8

Zeros are especially hard to understand. Constructing combinations such as 3 + 600 + 8000 helps show that the zero serves as a spacer.

			3
	6	0	0
8	0	0	0
8	6	0	3

The Source for Visual-Spatial Disorders   Copyright © 2002 LinguiSystems, Inc.

Mathematics: Number and Place Value, *continued*

Overview the number system: know the names of commas and use this chart to read and write large numbers.

Thousand	1,000
Million	1,000,000
Billion	1,000,000,000
Trillion	1,000,000,000,000
Quadrillion	1,000,000,000,000,000
Quintillion	1,000,000,000,000,000,000
Sextillion	1,000,000,000,000,000,000,000

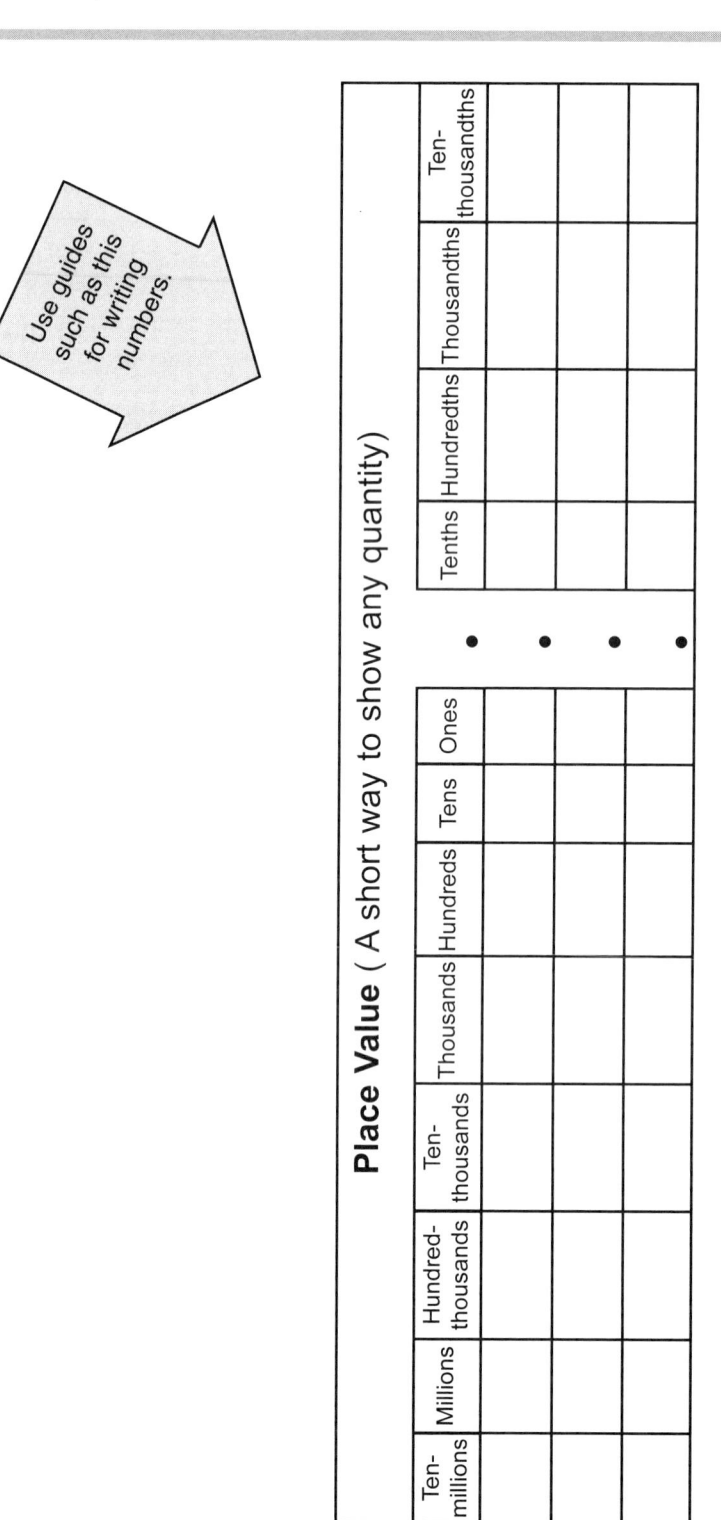

Use guides such as this for writing numbers.

Place Value (A short way to show any quantity)

Ten-millions	Millions	Hundred-thousands	Ten-thousands	Thousands	Hundreds	Tens	Ones	.	Tenths	Hundredths	Thousandths	Ten-thousandths
								.				
								.				
								.				

Reciprocal actions build understanding.

Mathematics: Operations

> For students who struggle with relative value and with sequences

Frame of Reference:
Students create their own overview of basic operations on one page.

Imagery:
Students create images for the actions within each operation.

Language:
Students talk someone else through each operation to test their scripts.

Students with visual-spatial disorders crave whole-to-parts instruction with references so they can keep track of the many subskills and compare operations. They also need their own language to translate actions into personal images. Create a personal summary for each operation in the student's own language on a single page. Have the student practice by using the script with a fellow student.

+ Addition

$$\begin{array}{r} 24 \\ +68 \\ \hline \end{array} \qquad \begin{array}{r} {}^{1} \\ 24 \\ +68 \\ \hline 92 \end{array}$$

1. Write the numbers lined up.
2. Add the numbers in the ones column first: 8 + 4 = 12. That is more than 9 so you put the 2 down and put the 1 above the next column. (You are really putting 10 there.)
3. Add the next column including the 1 you put up there: 6 + 2 + 1 = 9
4. The answer is 92.

— Subtraction

$$\begin{array}{r} 65 \\ -36 \\ \hline \end{array} \qquad \begin{array}{r} {}^{5}{}^{1} \\ \cancel{6}5 \\ -36 \\ \hline 29 \end{array}$$

1. Write the numbers so the one you are taking away is on the bottom.
2. Start at the ones column on the right. You want to take 6 from 5 but the 5 is too little.
3. So you go to the next column, make the 6 a 5 and bring the 10 over to the ones column and put it with the 5 to make 15.
4. Now you can take 6 from 15 = 9
5. You go to the next column and take 3 from the 5 = 2. The answer is 29.

The Source for Visual-Spatial Disorders

Mathematics: Operations, *continued*

---

x Multiplication

$$\begin{array}{r} 24 \\ \times\ 36 \\ \hline \end{array} \qquad \begin{array}{r} {\scriptstyle 1\,2} \\ 24 \\ \times\ 36 \\ \hline 144 \\ 72 \\ \hline 864 \end{array}$$

1. Pretend the problem is just 6 x 24. Multiply 6 x 4 = 24. Put the 4 under the line and put the 2 above the other 2. You are carrying it. Go 6 x 2 = 12 plus the 2 you carried = 14. Write it so the number reads 144.

2. Pretend the problem is 3 x 24. 3 x 4 = 12. Put the 2 down a line and under the 3 and carry the 1. 3 x 2 = 6 plus 1 = 7. This row reads 72.

3. Now add the two rows. The answer is 864.

÷ Division

$$\begin{array}{r} 187 \\ 8\overline{)1496} \\ \underline{8} \\ 69 \\ \underline{64} \\ 56 \\ \underline{56} \\ 00 \end{array}$$

| Dad says divide | Mom says multiply | Sister says subtract | Brother says bring down |

Follow the division family.

1. How many 8's are in 1496? Dad says divide 8 into 1; can't so divide into 14 = 1.
2. Mom says multiply. 1 x 8 = 8. Write that.
3. Sister says subtract. 14 - 8 = 6.
4. Brother says bring down the 9. Now back to Dad who says 8 into 69 = 8.
5. Mom says multiply 8 x 8 = 64. Sister says subtract. 69 - 64 = 5. Brother says bring down the 6.
6. Now it's back to Dad to divide 8 into 56 = 7.
7. Mom says multiply 7 x 8 = 56.
8. Sister says subtract. 56 - 56 = 0. There's nothing to bring down. Yeah!
9. The answer is 187.

Mathematics: Story Problems

For students who struggle with calculation, math operations, and the spatial imagery of language

Frame of Reference:
The student needs many references starting with number, and then place value, vocabulary, basic and advanced operations.

Imagery:
The student needs to image vocabulary as well as actions that have strong spatial features such as *combine*, *invert*, *distribute*, and *parallel*.

Language:
The student needs to accurately use the vocabulary of math to describe actions.

Story problems reveal learning disorders and expose gaps in instruction. The first and most complex stage of helping students who struggle with story problems is to sort out the reason or combination of reasons for their difficulties. Ideally, each cause is addressed and the student develops the underlying skills to solve story problems. However, students with fundamental learning disorders often have difficulty with story problems even after the subskills have been developed. Writing story problems is a valuable process to help students apply their newly-developed skills in an orderly progression. Writing demands the language of math which is the tool the students need to recall, integrate, and transfer their math knowledge.

Problem	Solution
• Reading (decoding) skills may be inadequate.	• Use a reader or have the problems tape recorded.
• Imaging math vocabulary may be deficient.	• Create visual notes for math vocabulary.
• Imaging events and actions may be deficient.	• Use objects and drawing to illustrate the events and actions.
• Sequencing skills may be insufficient.	• Show and record each step.
• Insufficient control of operations	• Review operations and/or use a calculator.
• Insufficient control of number and place value	• Review place value and provide guides for perceptual problems and directional confusion.
• Inability to question and self-guide with language	• Develop the student's language skills and teach scripting.

Mathematics: Story Problems, continued

Writing Story Problems:
Levels appropriate for a middle school student with a spatial disorder

Level One: The student uses real objects such as items from his desk. He composes addition, subtraction, multiplication, and division problems. Each problem is demonstrated in three dimensions, written in number form, solved, and explained.

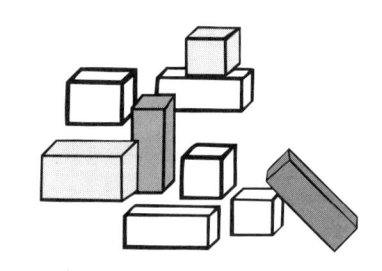

Level Two: The student uses blocks or some other object to represent real objects. She draws each problem, converts it to number form, solves it, and explains her solution.

Level Three: The student lists the items in the problems and draws events that require **2 operations** to solve. He then converts each problem to number form, solves it, and explains his solution.

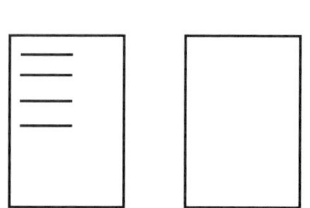

Writing

> *For students who struggle with written composition*

Many students, especially students with learning disabilities, don't realize that they know as much as they do. The information is stored in the vault of the student's brain, but she needs language to unlock that vault. A writing conference that relies on questioning will provide the key. You can conduct a writing conference with questioning to help students articulate what they do know.

When the student doesn't know how to begin

If you can take notes on what the student says, do so.

1. Begin with easy questions of fact.

    "What is the assignment?"
    "Who wrote _____?"

2. Move to more difficult statements of fact.

    "Who are the main characters?"
    "What happens in the book or story?"
    "What have you discovered in your research of the topic?"

Do not correct spelling, grammar, or punctuation until you have a very good draft.

3. Ask questions using *how*.

    "How does the main character know he has a problem?"
    "How does she get to the place where she's going?"

4. Ask questions using *why*.

    "Why did the main character choose to do what she did?"
    "Why did women in the 1920s choose to go to college?"

5. Ask questions that begin to focus the topic.

    "What do you find most interesting about what you've been telling me?" "Why?"

Writing, *continued*

The writing conference is ideally a conversation, a give-and-take in which you combine clarifying questions with the questions that move the discussion forward. Offer responses about what you find truly interesting in the conversation. Do not rush to be finished. It may take several writing conferences before the student is ready to sit down to write. Be patient.

When the student has generated plenty of information but doesn't know what to do with it

1. Ask questions that help focus the topic.
   "What is the most interesting thing you've learned about ____?"
   "What issues or points are still unresolved for you?"
   "What is the author's point?" "Do you agree with it?"

2. Ask questions about patterns or connections.
   "Do you see a pattern in the information?"
   "What themes seem to be running through the information?"

3. Ask questions that lead to the thesis or main point.
   "What do you want to say about ____?"

4. Ask questions that lead to the writing of an introduction.
   "What would a reader need to know in order to follow what you're going to say?"
   "What would readers need to understand in order for them to be prepared for the thesis or main point?"

5. Ask questions that lead to the rest of the paper.
   "Given your main point, what is the next thing the reader needs to know to understand your point?" Nudge the student toward a preliminary outline from the information recorded by making connections and asking questions about what connections the student sees.

6. If a student is having trouble getting words on paper, offer to do the drafting while the student dictates. Use her exact words but don't hesitate to probe for clarification. Once a draft is complete, repeat the series of questions, asking the student to fill in gaps rather than just trying to add more at the end.

7. For the conclusion, ask again, "What do you think this means?"

Selected Bibliography: Spatial Relationships

Arnheim, R. 1969. *Visual Thinking*. Los Angeles: University of California Press. The processes of vision involve thinking and reasoning. Perception, in Arnheim's view, is the stuff of thinking through which we structure events and derive ideas. "The language of images is a prime mover of the constructive, creative imagination. Thinking calls for images and images contain thought." Dr. Arnheim was Professor Emeritus of Psychology of Art at Harvard at the time he wrote this book.

Barlow, H., C. Blakemore, and M. Weston-Smith, eds. 1990. *Images and Understanding*. Cambridge, MA: Cambridge University Press. The editors present a comprehensive collection of papers on the form, function, and processes involved with imaging. They convey how images help make meaning and participate in thought.

Bloom, P., M. A. Peterson, L. Nadel, and M. F. Garrett, eds. 1996. *Language and Space*. Cambridge, MA.: The MIT Press. The book emanated from a conference that sought to answer questions such as "How do we represent space?" and "What role does culture play in spatial representation?" Behind these investigations is the concern of how to better understand orientation and left-right confusion.

Bloomer, C. M. 1976. *Principles of Visual Perception*. New York: Van Nostrand Reinhold Company. The author explains and illustrates the mechanisms of visual perception.

Dondis, D. A. 1973. *A Primer of Visual Literacy*. Cambridge, MA: The MIT Press. The author describes the essential skills needed to create and to understand visual communication.

Epstein, W. and S. Rogers, eds. 1995. *Perception of Space and Motion*. San Diego: Academic Press. The authors present current information about the application of perception in a variety of situations that include depth, pictorial layout, events, 3-dimensional structure as well as the development of space and motor perception.

Gombrich, E. H. 1969. *Art and Illusion*. Princeton, NJ: Princeton University Press. The author originally presented these ideas at the A. W. Mellon Lectures in Fine Arts at Princeton. The theme of the work is "A Study in the Psychology of Pictorial Representation."

Jackendoff, R. S. 1993. *Languages of the Mind: Essays on Mental Representation*. Cambridge, MA.: The MIT Press. These essays investigate mental representation of widely different content including spatial information.

Kosslyn, S. M. 1983. *Ghosts in the Mind's Machine*. New York: W. W. Norton & Company. The author explores how humans create and use images in the brain.

Kosslyn, S. M. 1980. *Image and Mind*. Cambridge, MA: Harvard University Press. This author explores imagery, how images are created and used, and makes a case for the central role of images in the function of the mind.

McKim, R. H. 1972. *Experiences in Visual Thinking*. Monterey, CA: Brooks/Cole Publishing Company. This is a Stanford University textbook used in the course on visual thinking developed by the author.

Selected Bibliography, continued

McKim, R. H. 1980. *Thinking Visually*. Belmont, CA: Lifetime Learning Publications. This strategy manual for problem solving includes a wide range of exercises to help individuals learn to solve visual problems.

Neff-Lippman, Julie. 1988. "Working with Learning Disabled Students in the Writing Center," in Irene Clark, *Writing in the Center*, Third Edition. Dubuque, IA: Kendall/Hunt Publishing Company. This chapter provides detailed instructions for those who work with learning disabled students one-to-one.

Neff-Lippman, Julie. 1994. "The Writing Center and the Learning Disabled Student," in Joan Mullin and Ray Wallace (eds.), *Intersections*. Urbana, IL: NCTE. This chapter, in a book about writing center pedagogy and theory, discusses how research about learning disabilities informs practice in the writing center.

Pick, A. D., ed. 1979. *Perception and Its Development: A Tribute to Eleanor J. Gibson*. Hillsdale, NJ: Lawrence Erlbaum Associates, Publishers. The authors present a series of papers that portray the nature and development of perception.

Potegal, M., ed. 1982. *Spatial Abilities, Development, and Physiological Foundations*. New York: Academic Press. The authors examine the way organisms learn to understand spatial relationships and review the physiological and neurological mechanisms involved.

Sacks, O. 1984. *A Leg to Stand On*. New York: Harper & Row, Publishers. The author, who is a neurologist, probes self awareness and consciousness in a personal episode that occurred after a severe leg injury. During convalescence he went through an extended interval of feeling dissociated from his leg even though it had healed.

Samuels, M., M.D. and N. Samuels. 1980. *Seeing with the Mind's Eye*. New York: Random House. This comprehensive book explores the history, techniques, and uses of visualization.

Schone, H. 1984. *Spatial Orientation*. Princeton, NJ: Princeton University Press. Subtitled "The Spatial Control of Behavior in Animals and Man," this book describes how animals orient themselves in space as well as the mechanisms and processes involved.

Scientific American Readings. 1972. *Perception: Mechanisms and Models*. San Francisco: W. H. Freeman and Company. This collection of articles includes fundamental research papers on perception, systems, and processes.

Selected Bibliography, continued

Journals

Silbert, A., P. Wolff, and J. Lilienthal. 1977. "Spatial and Temporal Processing in Patients with Turner's Syndrome." *Behavioral Genetics* 7, no. 1:11-21.

Money, J. 1964. "Two Cytogenetic Syndromes: Psychological Comparisons—Intelligence and Specific Factor Quotients." *Journal of Psychiatric Research* 2:223-231

Garron, D. C. 1977. "Intelligence Among Persons with Turner's Syndrome." *Behavioral Genetics* 7:105-127

Internet

ARK Foundation. The video designed to accompany *The Learning Window Handbook* is available by contacting ARKFdn@aol.com.

Stockdale, C. and Possin, C. *Spatial Relations and Learning.*
<http://www.newhorizons.org/spneeds_arkspatial.html>

Stockdale, C. and Possin, C. *The Learning Window Handbook.*
<http://www.newhorizons.org/spneeds_arkwindow.html>

Turner Syndrome Society of the United States. <http://www.turner-syndrome-us.org>

Entering "basic training" for high school — a lot is unknown
sequencing activities & making choices to plan
pacing → realistic time allotments

real stress / perceived stress / how to keep
    stress at bay

visual imaging what is read is often very difficult.
    discussion + prediction helps

Spatial = comparisons & relationships

T:
    Help students not use their disabilities
    as an excuse.
    not working: stop doing it!
    don't want to be singled out; however,
        need accommodations